D1175591

Introduction to
QUANTUM MECHANICS

Introduction to
QUANTUM
MECHANICS
THIRD EDITION

Paul T. Matthews, F.R.S.

London · New York · St Louis · San Francisco · Düsseldorf
Johannesburg · Kuala Lumpur · Mexico · Montreal · New Delhi
Panama · Paris · São Paulo · Singapore · Sydney · Toronto

Published by
McGRAW-HILL Book Company (UK) Limited
MAIDENHEAD·BERKSHIRE·ENGLAND

07 084036 9

PRINTED AND BOUND IN GREAT BRITAIN

To MARGIT

to whom all the great ideas in physics are
obvious, in the hope that this book will help
others to agree with her.

PREFACE TO THE THIRD EDITION

As stated in the Preface to the first edition, this book was originally written very specifically as an introduction to more complete texts. For this reason no approximate methods were included. However, quantum mechanics now figures much more widely in degree courses than it did ten years ago, and there are many students who want a general understanding, without the detailed knowledge necessary for those who wish to specialize in some branch of physics in which quantum mechanics is essential. To make this text adequate in itself to meet this demand, a substantial extra chapter has been added introducing time-dependent perturbation theory and applying it to transitions and scattering cross-sections. It is applied explicitly to Rutherford scattering, radiative transitions, and weak nuclear interactions (β-decay), thereby providing a full theoretical basis for three very important topics, previously covered only by general statements.

To comply with modern practice, the text has also been converted to MKS units. This is achieved very largely by the device of introducing

$$e_M^2 \equiv e^2/(4\pi\epsilon_0),$$

which avoids numerous factors of $4\pi\epsilon_0$ which otherwise obscure the discussion, particularly of the hydrogen atom.

P. T. MATTHEWS

PREFACE

This book is based on a course of lectures, which has been given for a number of years to physics students. As the title implies, it is written for people who have not previously studied quantum mechanics. The emphasis is on concepts and the mathematical machinery has been kept to a minimum. The reader is never assumed to know how to do anything more complicated than differentiate the product of two functions. On the other hand, the introduction to new physical ideas is based on an attempt to get right to the heart of the matter from the start. Thus much less emphasis than usual is put on wave functions, and the Schrödinger equation turns up as a special case of the eigenvalue equations which determine the possible values of any quantum observable.

The development of the new physical ideas is covered in the first five chapters, which constitute Part I. The notion of operators is introduced, but only differential operators are used. Matrices appear for the first time in connection with spin in § 8.2–§ 8.3 and are introduced using the Dirac notation. The development of the general theory is taken up again in Part IV. The Dirac notation is there introduced much more fully, in particular in Chapter 12, and is related to the wave mechanical formalism. This notation enormously clarifies the mathematical structure of the theory, and Dirac's *Quantum Mechanics* (Oxford University Press, 4th edition, 1958), with its penetrating depth and broad sweep, seems destined to remain indefinitely the classic treatment of this subject. If this text helps students to grasp more easily the Dirac formulation, it will have served its main purpose.

Between the two parts of the book which deal with general quantum theory, and could be read consecutively, are two parts devoted to specific applications.

Part II is about atomic theory. Here again the emphasis is entirely on concepts. Chapters 6 and 7 deal in detail with the derivation of the eigenvalues of orbital angular momentum and the Schrödinger formula for the level structure of the hydrogen atom, because these provide the most conclusive evidence that the theory set up in Part I is correct. Chapter 8 introduces the notions of spin and statistics.

Later developments in the theory of the hydrogen levels, arising from the Dirac equation and the Lamb shift, are sketched in the final section (§ 8.6).

Part III considers nuclear theory in a purely introductory fashion. The discussion centres about the essential problem of investigating the nucleon–nucleon potential by means of scattering experiments and Chapter 10 consists entirely of an introduction to the theory of scattering. Here again the way in which this subject has led the search for fundamental laws into High Energy (or Elementary Particle) physics is outlined in the final section (§ 11.4).

In Parts II and III our object has been to exhibit the wood rather than the trees. For this reason we have not introduced any of the standard, and important, approximate techniques for calculating either energy levels or scattering cross-sections. These are dealt with in a number of well-established works such as those of Landau and Lifshitz (L. D. Landau and E. M. Lifshitz, *Quantum Mechanics*, Pergamon Press, 1959), and Schiff (L. I. Schiff, *Quantum Mechanics*, McGraw-Hill, 2nd edition, 1955), and it is hoped that this text will also serve as an introduction to larger books of this type.

P. T. MATTHEWS

CONTENTS

PART I

Basic Formulation

CHAPTER 1

INTRODUCTION

Quantum mechanics is the theory of atomic and nuclear systems. It has been developed out of classical physics, particularly the two main branches, Newtonian mechanics and Maxwell's electromagnetic theory. We start by outlining the concepts of classical theory. We then show how these concepts proved quite inadequate to describe atomic systems, and state the rules of thumb which were superposed on classical theory by Planck, Bohr, and de Broglie, to make up what is usually called Old Quantum Theory. This provided a philosophically very unsatisfactory, but at least partially successful, description of atomic phenomena. It led the way to the fundamental reformulation of the physical theory of microscopic systems, which we develop in the subsequent chapters.

§ 1.1 Classical Physics

(a) *Newtonian Mechanics*

In classical physics matter is regarded as being made up of point particles. They move under the action of mutually interacting forces, according to Newton's laws. The most important of these is the law of motion,

$$\text{Force} = \text{mass} \times \text{acceleration}.$$

Combined with the law of gravity, this theory was triumphantly successful in explaining the motion of the planets, and provided a satisfactory description of the motion of electrically neutral macroscopic systems, generally.

The essential point is that matter is treated in terms of particles of definite mass, the motion of a free particle being defined in terms of its energy, E, and momentum, $\mathbf{p}$.

(b) *Electromagnetic Theory*

The other main body of classical physics is concerned with electric and magnetic phenomena, which are best described in terms of electric and magnetic fields, $\mathscr{E}(x)$ and $\mathbf{B}(x)$, respectively. These are related to charge and current densities by Maxwell's equations, which we will not repeat, since they can be found in any standard text, and we do not need them specifically. The key point for our purpose is that they lead to the conclusion that, in free space, both the electric and magnetic fields satisfy the equation

$$\left(\frac{1}{c^2}\frac{\partial^2}{\partial t^2}-\nabla^2\right)\begin{Bmatrix}\mathscr{E}(x)\\ B(x)\end{Bmatrix} = 0. \tag{1.1}$$

This states that these fields propagate through space as waves with a constant velocity, c. It was Maxwell's inspired guess that these waves, for the appropriate frequencies, should be identified with visible light. We have since become familiar with other forms of such radiation, from the extremely low frequency bands used in radar and radio astronomy, through the visible range, to the high frequency radiation of X-ray photography and the γ-rays of radio-active fall out.

It is well known from geometrical optics that there are many properties of radiation for which the wave concept is not essential. However, the interference phenomena, of which diffraction is typical, definitely require the wave picture for a satisfactory explanation. The typical wave is expressed by

$$\exp\left[-i(\omega t-\mathbf{k}.\mathbf{x})\right], \tag{1.2}$$

where the angular frequency ω, and the propagation vector $\mathbf{k}$ are the basic physical characteristics of the wave.† Since the velocity is c,

$$\omega = |k|c. \tag{1.3}$$

These two disciplines of mechanics and electromagnetism are coupled by the Lorentz law, which states that a particle of charge e, moving in electric and magnetic fields with velocity $\mathbf{v}$, is subject to a force

$$\mathbf{F}(x) = e(\mathscr{E}(x)+\mathbf{v}\wedge B(x)). \tag{1.4}$$

In principle this classical picture of the world, with matter consisting of point particles, and radiation consisting of waves, could have provided the framework for a fundamental description of all physical phenomena; the point particles being protons and electrons, each carrying mass and unit electric charge, and interacting through the fundamental electromagnetic and gravitational forces. However, even before the discovery of the proton, classical concepts were proving quite inadequate to describe either the motion of electrons or their interaction with radiation.

† Related to frequency ν and wave-length λ by $\omega = 2\pi\nu$, $|k| = 2\pi/\lambda$.

§ 1.2 Breakdown of Classical Concepts and Old Quantum Theory

(a) *Particle Aspects of Radiation and Planck's Hypothesis*

Historically the first indication of a breakdown of classical concepts occurred in the rather complicated phenomenon of "black body" radiation, which is concerned with the thermodynamics of the exchange of energy between radiation and matter. Classically it is assumed that this exchange of energy is continuous, in the sense that light of frequency ω can give up any amount of energy on absorption, the precise amount in any particular case depending on the energy intensity in the beam. It was shown by Planck† that a correct thermodynamic formula is obtained only if it is assumed that the energy exchange is discrete. Specifically Planck postulated that radiation of angular frequency ω can only exchange energy with matter in units of $\hbar\omega$, where $\hbar$ is a universal constant. (Planck's constant is

$$h = 2\pi\hbar = 6\cdot62 \times 10^{-34} \text{ joule sec)} \qquad (1.5)$$

Planck's hypothesis may be stated by saying that the radiation of frequency ω behaves like a stream of particles (photons) of energy

$$E = \hbar\omega, \qquad (1.6)$$

which may be emitted or absorbed by matter. Since they travel with the velocity of light, according to the theory of Special Relativity, their rest mass must be zero. The relativistic relation between energy and momentum,

$$E^2/c^2 = p^2 + m^2 c^2, \qquad (1.7)$$

then shows that for photons $(m = 0)$

$$p = E/c. \qquad (1.8)$$

The equations (1.3), (1.6), and (1.8) can be combined, by eliminating c, into

$$\boxed{\begin{aligned} E &= \hbar\omega, \\ \mathbf{p} &= \hbar\mathbf{k}. \end{aligned}} \qquad (1.9)$$

This shows clearly the relation between the particle parameters $(E, \mathbf{p})$ of the photon, and the parameters $(\omega, \mathbf{k})$ of the corresponding wave.

A much simpler example of the particle aspect of radiation is the photo-electric effect. If a beam of monochromatic light, frequency ω,

† M. Planck, *Ann. Physik*, **4**, 553 (1901).

is played on the surface of a metal, electrons may be emitted. If $\hbar\omega$ is less than some limit W, which depends on the particular metal, no electrons are emitted over a wide range of intensities of the beam. If $\hbar\omega > W$, electrons are emitted with kinetic energy T, where

$$\hbar\omega = W + T. \tag{1.10}$$

Note that even when electrons are emitted, their energy T does not depend on the intensity of the radiation, but only on its frequency. This is completely unintelligible on the classical picture of a continuous exchange of energy, as defined above. However, it is quite simple to understand on Planck's photon hypothesis. The quantity

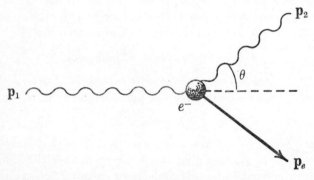

FIG. 1.1. Diagram of the collision of a photon, momentum $\mathbf{p}_1$, with a stationary electron, in which the electron recoils with momentum $\mathbf{p}_e$ and the photon with momentum $\mathbf{p}_2$.

W is the work which is required to free the electron from the attractive potential produced by the metal. The energy $\hbar\omega$ is transmitted by the photons and, if the photon energy is less than W, no electrons are emitted. If it is greater than W, when a photon gives up its energy $\hbar\omega$ to an electron, the electron escapes with kinetic energy given by (1.10).

The photo-electric effect is a very specific confirmation of Planck's hypothesis, since it is directly controlled by the mechanism of exchange of energy between radiation and electrons, and no other physical effects are involved.

The photo-electric effect, and black body radiation show only that energy exchange takes place by quanta $\hbar\omega$. The particle nature of the radiation itself is shown most clearly in the scattering of X-rays by electrons (Compton effect). Consider a photon of momentum $\mathbf{p}_1$

(energy $p_1 c$), colliding with a stationary electron. After the collision the photon has momentum $\mathbf{p}_2$ (energy $p_2 c$), and the electron (mass m) has momentum $\mathbf{p}_e$. By momentum conservation,

$$\mathbf{p}_1 = \mathbf{p}_2 + \mathbf{p}_e. \tag{1.11}$$

Therefore
$$p_e^2 = p_1^2 + p_2^2 - 2p_1 p_2 \cos \theta. \tag{1.12}$$

By (relativistic) energy conservation (see (1.7))

$$p_1 + mc = p_2 + (p_e^2 + m^2 c^2)^{1/2}. \tag{1.13}$$

Eliminating p_e^2 from (1.13) and (1.12) gives, very simply,

$$mc(p_1 - p_2) = 2p_2 p_1 \sin^2 \theta/2. \tag{1.14}$$

Dividing by $p_2 p_1$, and expressing the result in terms of wave-length, which by (1.9) is

$$\lambda = \frac{\hbar}{p}, \tag{1.15}$$

we obtain

$$\lambda_2 - \lambda_1 = 2\lambda_e \sin^2 \theta/2. \tag{1.16}$$

Here λ_e is the electron Compton wave-length,

$$\lambda_e = \frac{\hbar}{mc} \simeq 4 \times 10^{-13} \text{ m}. \tag{1.17}$$

This change in wave-length, which depends only on the angle of scatter of the radiation and not on the original frequency, is precisely what is observed. It comes directly out of the simple particle picture of the collision of a photon with an electron, but cannot be explained in terms of radiation waves.

(b) Wave Aspects of Matter and de Broglie's Hypothesis

Complementary to the above effects which revealed the particle aspects of radiation, were the experiments of Davisson and Germer,[†] who showed that a beam of electrons reflected from the surface of a nickel crystal form diffraction patterns, exactly analogous to the diffraction of light by a grating. The pattern persists even when the intensity of the electrons is so low that they pass through the apparatus one at a time. Diffraction is essentially a wave phenomenon, and its appearance under these circumstances shows that a wave of the general form (1.2) must in some way be associated with the motion of a single electron, normally described by the particle parameters E and $\mathbf{p}$.

[†] C. Davisson and L. Germer, *Nature*, **119**, 558 (1927).

Even before the Davisson–Germer experiments, it had been suggested by de Broglie† that the formula (1.9) which relates the particle and wave aspects of radiation should apply also to electrons. Thus an electron of given energy and momentum was associated in some rather ill-defined way with a de Broglie wave

$$\exp[-i(Et-\mathbf{p}.\mathbf{x})/\hbar].\qquad(1.18)$$

This relation between wave and particle parameters, with the value of $\hbar$ already determined by radiation effects, gives correctly the observed relation between the width of the diffraction bands, and the energy of the electrons (see Problem **1.5**).

(c) Discrete Levels and the Bohr Hypothesis

The failure of classical concepts when applied to the motion of electrons appeared most clearly in connection with the hydrogen atom. The experiments of Rutherford‡ showed that an atom can be regarded as a negatively charged electron orbiting around a relatively massive, positively charged nucleus (for hydrogen a single proton). With the neglect of radiation, this system is exactly analogous to the motion of a planet round the sun, with the gravitational attraction between the masses being replaced by the Coulomb attraction between the charges. It is ironical that the electrical analogue of Newton's gravitational triumph should provide the greatest debacle for classical theory. The reason is, of course, that the radiation cannot be neglected. The orbiting electron constitutes a rapidly accelerating charge, which according to Maxwell's theory acts as a source of radiant energy. According to classical theory, in a matter of 10^{-10} sec the electron should coalesce with the proton, giving off its mechanical energy in the form of a brief flash of light.

The frequency of the emitted radiation is related to the frequency of the electron in its orbit. As the electron radiates energy, this frequency, again according to classical theory, should change rapidly but continuously, thus giving rise to radiation with a continuous range of frequencies.

The classical theory of the Rutherford atom thus has two important qualitative features.

(i) The atom should be very unstable.

(ii) It should radiate energy over a continuous range of frequencies.

Both these results are completely contradicted by experiment. The first very obviously, since the atoms are about the most stable objects

---

† L. de Broglie, *Ann. Phys.* **3**, 22 (1925).
‡ E. Rutherford, *Phil. Mag.* **21**, 669 (1911).

we know. (If they do break up, it is normally by ionization, in which the electrons are separated from the nuclei, which is the reverse effect to the classical instability.) The second effect is less obviously incorrect, but a detailed study of the radiation from hydrogen by Balmer, as early as 1885, showed that the emitted frequencies were discrete, and that some of the most easily observed lines satisfied the empirical relation,

$$\omega = N\left(\frac{1}{2^2} - \frac{1}{n^2}\right), \qquad n = 3, 4, 5, \ldots \tag{1.19}$$

This observation of a classically continuous variable, appearing physically with only a discrete set of possible values, is the crucial, qualitatively new, feature of the atom.

Some rules of thumb, which enable one to extract the observed results from a semi-classical theory, were proposed by Bohr.† For simplicity we will state these for the case of circular orbits. Then Bohr's rules are the following:

(i) The magnitude of the angular momentum l of the electron is an integer multiple of $\hbar$.

$$l = n\hbar, \qquad n = 1, 2, \ldots \tag{1.20}$$

This requirement of discrete values for the angular momentum immediately leads to discrete values of the energy E_n.

(ii) The radiation occurs when the electron makes discontinuous jumps from an orbit of energy E_n to one of energy $E_{n'}$, say, and the resulting angular frequency $\omega_{nn'}$ is determined by

$$\hbar\omega_{nn'} = |E_n - E_{n'}|. \tag{1.21}$$

We now apply these rules to the hydrogen atom whose electron, mass m, revolves around a nucleus (taken as fixed) in a circular orbit of radius a, with angular velocity w. Then the equation of motion, which relates the Coulomb attraction to the radial acceleration, is‡

$$\frac{e_M^2}{a^2} = maw^2. \tag{1.22}$$

Bohr's condition (i) is

$$ma^2w = n\hbar, \qquad n = 1, 2, \ldots \tag{1.23}$$

Solving these gives a discrete set of possible radii,

$$a_n = \left(\frac{\hbar^2}{me_M^2}\right)n^2 \equiv a_0 n^2, \tag{1.24}$$

† N. Bohr, *Phil. Mag.* **26**, 476 (1913).

‡ All electromagnetic quantities are expressed in mks units and we have introduced:

$$e_M^2 \equiv e^2/(4\pi\epsilon_0).$$

and
$$w = \frac{me_M^4}{\hbar^3}\frac{1}{n^3}. \tag{1.25}$$

The energy is made up of the kinetic and potential parts. Thus the discrete energy levels are

$$E_n = \frac{m}{2}a_n^2 w^2 - \frac{e_M^2}{a_n} = -\frac{me_M^4}{2\hbar^2}\frac{1}{n^2}$$

$$= \left(-\frac{1}{2}\frac{e_M^2}{a_0}\right)\frac{1}{n^2}. \tag{1.26}$$

The fundamental length a_0, defined by (1.24), is the radius of the lowest Bohr orbit. According to (1.21) the possible angular frequencies of radiation are

$$\omega_{nn'} = \frac{e_M^2}{2\hbar a_0}\left(\frac{1}{n^2} - \frac{1}{n'^2}\right). \tag{1.27}$$

The series with $n = 2$ and $n' = 3, 4, 5$ has four lines which lie in the visible region and form part of the Balmer series quoted above.

The Bohr atom is stable, since no further radiation of energy is possible once it has reached the lowest level, E_1, given by (1.26).

§ 1.3 Summary

The procedure of explaining atomic phenomena in classical terms by the superposition of additional arbitrary relations was carried further than has been indicated above, but there is no purpose in giving more details. It is clearly intrinsically unsatisfactory to have both radiation and matter being treated sometimes as waves and sometimes as particles in an apparently arbitrary manner, and discrete hydrogen levels being produced by *ad hoc* rules, which are completely contrary to the spirit of classical mechanics, to which they are applied. What is required is a basic reformulation of the theory in such a way that both those classical concepts which remain correct, and the Planck–Bohr–de Broglie rules, shall appear as natural consequences of some coherent whole. This is the quantum theory, which we set up in Chapter 3. For this purpose the above rules are an important guide. The essentially non-classical features which need to appear naturally out of the new theory are:

(i) the particle aspect of radiation—photons—(Planck);

(ii) the wave aspect of particles—(de Broglie);

(iii) some physical variables having a discrete set, rather than a continuous range, of possible values—in particular the hydrogen levels—(Bohr).

PROBLEMS I

(For the values of physical constants see Appendix.)

1.1. The angular frequency ω of visible light is about 10^{16} sec^{-1}. This can be equated to a typical Bohr transition as determined by (1.27). Taking e and m as known, show that this implies an order of magnitude of $\hbar$ consistent with that found by Planck, $(\hbar \sim 10^{-34}$ mks).

1.2. What is the radius of the lowest Bohr orbit in cm? What is the energy of the ground state in ergs and in electron volts? Show that the kinetic energy in any circular Coulomb orbit is equal in magnitude, but opposite in sign, to the total energy. Hence, show that the tangential momentum in the lowest Bohr orbit is $\sim 2 \times 10^{-24}$ mks.

1.3. Find the time taken per cycle in the lowest Bohr orbit in terms of e, $\hbar$ and m, and in seconds.

1.4. Show that velocity v of the electron in the lowest Bohr orbit is $e_M^2/\hbar$ and express this as a fraction of the velocity of light, c. (This ratio $v/c = e_M^2/(\hbar c)$ is known as the fine structure constant. See (8.60) and § 14.3.)

Is it reasonable, as a first approximation, to neglect relativistic effects in the theory of the hydrogen atom?

1.5. Show that the wave-length, λ, associated with an electron of kinetic energy, T, according to the de Broglie formula (1.9), is

$$\lambda = \frac{2\pi\hbar}{\sqrt{(2mT)}}.$$

By the standard diffraction grating formula, the wave-lengths, which interfere constructively in the deflection of electrons at an angle θ from the surface of a crystal, with inter-atomic spacing d, are

$$n\lambda = d\sin\theta.$$

For

$$n = 1, \quad \sin\theta \simeq 1 \quad \text{and} \quad d \simeq a_0,$$

we have

$$\lambda \simeq a_0.$$

Equating the two expressions for λ, we obtain an approximate formula for $\hbar$, in terms of the kinetic energy of the electrons in a diffraction experiment,

$$2\pi\hbar = \sqrt{(2mT)}\,a_0.$$

Show that $T \simeq 50$ eV (the value in the Davisson–Germer experiment) implies a value of $\hbar$ consistent with that obtained by Planck and from the Bohr formula.

CHAPTER 2

OPERATORS

§ 2.1 Definitions and Operator Equations

Before proceeding to the setting up of quantum mechanics, we now develop that part of the mathematical theory of operators which plays an essential role in the later work.

Loosely speaking, an operator, which we denote by $\hat{A}$, is any mathematical entity which operates on any function of x, say, and turns it into another function.† The simplest example of an operator is to take $\hat{A}$ itself to be a function of x, $\hat{A}(x)$, the operation being multiplication. Thus we might have

$$\hat{A}(x) = x.$$

The operator x operating on any function $\psi(x)$ produces the new function $x\psi(x)$.

A less trivial example is the operation of differentiation, the operator $\hat{A}$ being any function of $\partial/\partial x$,

$$\hat{A}\left(\frac{\partial}{\partial x}\right), \quad \left(\text{for example, } \hat{A} = \frac{\partial^2}{\partial x^2}\right)$$

The most general operator we need consider for the moment is a function of x and $\partial/\partial x$,

$$\hat{A} = \hat{A}\left(x, \frac{\partial}{\partial x}\right).$$

We now introduce the notion of an operator equation. Consider the operator

$$\hat{A}\left(x, \frac{\partial}{\partial x}\right) = \frac{\partial}{\partial x} x. \tag{2.1}$$

Then, for *any* function $\psi(x)$,

$$\left(\frac{\partial}{\partial x} x\right)\psi(x) = \psi(x) + x \frac{\partial\psi(x)}{\partial x} \tag{2.2}$$

$$= \left(1 + x\frac{\partial}{\partial x}\right)\psi(x). \tag{2.3}$$

† Throughout the book, operators are distinguished by such an accent—thus $\hat{O}$.

In obtaining the first equality (2.2) we have used the ordinary rule for the differentiation of the product $x\psi(x)$, the convention being the usual one that the derivative $\partial/\partial x$ operates on any function occurring to the right of it. Since the final equality (2.3) is valid for *any* $\psi(x)$, we can formally cancel the factor $\psi(x)$ on the right, and write the operator equation,

$$\frac{\partial}{\partial x} x = 1 + x \frac{\partial}{\partial x}. \tag{2.4}$$

In general, an operator equation,

$$\hat{A}\left(x, \frac{\partial}{\partial x}\right) = \hat{B}\left(x, \frac{\partial}{\partial x}\right) + \hat{C}\left(x, \frac{\partial}{\partial x}\right), \tag{2.5}$$

implies

$$\left[\hat{A}\left(x, \frac{\partial}{\partial x}\right)\right]\psi(x) = \left[\hat{B}\left(x, \frac{\partial}{\partial x}\right) + \hat{C}\left(x, \frac{\partial}{\partial x}\right)\right]\psi(x) \tag{2.6}$$

for *any* $\psi(x)$.

§ 2.2 The Eigenvalue Equation

To each operator $\hat{A}(x, \partial/\partial x)$ belong a set of numbers, a_n, and a set of functions, $u_n(x)$, defined by the equation,

$$\hat{A}\left(x, \frac{\partial}{\partial x}\right) u_n(x) = a_n u_n(x), \tag{2.7}$$

where a_n is an *eigenvalue*, and $u_n(x)$ is the corresponding *eigenfunction*. The eigenfunctions of an operator are thus those special functions which remain unaltered under the operation of the operator, apart from multiplication by the eigenvalue. Equation (2.7) is the eigenvalue equation of the operator $\hat{A}$. As an example consider

$$\hat{A}\left(x, \frac{\partial}{\partial x}\right) = -i\frac{\partial}{\partial x}, \tag{2.8}$$

with the boundary condition that $u_n(x)$ is periodic in the range L. Then the eigenvalue equation (2.7) is

$$-i\frac{\partial}{\partial x} u_n(x) = a_n u_n(x). \tag{2.9}$$

Thus

$$u_n(x) = e^{ia_n x} \tag{2.10}$$

where, by the boundary condition, a_n form the discrete set

$$a_n = \frac{2n\pi}{L}. \tag{2.11}$$

In the limit $L \to \infty$, the gap between successive eigenvalues tends to zero, and, going back to (2.9), it can be seen that the eigenfunctions go over into

$$u_a(x) = e^{iax}, \tag{2.12}$$

where the eigenvalue a becomes a continuous variable, which can take on any value. Note that the eigenvalues depend on the boundary condition imposed on the solutions to the eigenvalue equation, (2.7), so they are only well defined when this boundary condition is given.

§ 2.3 Commutation Relations

Finally, we consider the successive operation of two operators. We define the commutator of two operators $\hat{A}$ and $\hat{B}$ to be

$$[\hat{A}, \hat{B}] \equiv \hat{A}\hat{B} - \hat{B}\hat{A}, \tag{2.13}$$

which is the difference between operating first with $\hat{B}$ and then with $\hat{A}$, and first with $\hat{A}$ and then with $\hat{B}$. In general this is not zero,

$$[\hat{A}, \hat{B}] \neq 0, \tag{2.14}$$

but is some new operator.

To show this, it is sufficient to consider the simple case

$$\hat{A} = x. \qquad \hat{B} = \frac{\partial}{\partial x}. \tag{2.15}$$

Then, for any $\psi(x)$,

$$\left[x, \frac{\partial}{\partial x} \right] \psi(x) = \left(x \frac{\partial}{\partial x} - \frac{\partial}{\partial x} x \right) \psi(x)$$

$$= \left(x \frac{\partial}{\partial x} - 1 - x \frac{\partial}{\partial x} \right) \psi(x), \tag{2.16}$$

where the "one" in the bracket comes, as in (2.3), from the operation of $\partial/\partial x$ on the factor x which lies to its right. Since this is true for any $\psi(x)$, we have the operator equation

$$\left[x, \frac{\partial}{\partial x} \right] = -1. \tag{2.17}$$

An equation which determines the commutator of two operators is called a commutation relation. The special cases, such as (2.17), in which the commutator of two operators is a number, play a particularly important role in the subsequent theory.

§ 2.4 Summary

We have collected together the mathematical machinery which will be required in the next chapter. It consists of a few definitions. There is no more complicated mathematical operation involved than the differentiation of a product. However, there are some new ideas here. Before starting on the next chapter, which contains all the new physical concepts necessary for a basic theory of atomic systems, the reader is strongly recommended to make himself familiar with the mathematical notions introduced above.

We summarize the main properties of operators:

(i) With each operator is associated a set of numbers, the eigenvalues, through the eigenvalue equation (2.7).

(ii) In general the commutator of two operators is not zero;

$$[\hat{A}, \hat{B}] \equiv \hat{A}\hat{B} - \hat{B}\hat{A} \neq 0.$$

PROBLEMS II

2.1. Establish the operator equation

$$\frac{\partial}{\partial x} x^n = n x^{n-1} + x^n \frac{\partial}{\partial x}$$

and hence show that

$$\left[\frac{\partial}{\partial x}, x^n \right] = n x^{n-1}$$

2.2. Evaluate

$$\left[\frac{\partial}{\partial x}, \frac{\partial^n}{\partial x^n} \right]$$

2.3. Show that $u(x) = e^{-(1/2)x^2}$ is an eigenfunction of the operator

$$\hat{A}\left(x, \frac{\partial}{\partial x} \right) = \left(\frac{\partial^2}{\partial x^2} - x^2 \right).$$

and find the corresponding eigenvalue.

2.4. Verify the operator equations:

$$\left(\frac{\partial}{\partial x} + x \right)\left(\frac{\partial}{\partial x} - x \right) = \frac{\partial^2}{\partial x^2} - x^2 - 1,$$

$$\left(\frac{\partial}{\partial x} - x \right)\left(\frac{\partial}{\partial x} + x \right) = \frac{\partial^2}{\partial x^2} - x^2 + 1.$$

CHAPTER 3

QUANTUM MECHANICS

§ 3.1 Operations of Observation

Classical mechanics breaks down when applied to small systems, provided they are small enough. It can be applied satisfactorily to the stars in their courses and to the motion of a golf ball, but fails completely when applied to atoms. It would appear that when we get down to atomic systems the word "small" takes on an absolute, not just a relative, meaning. The understanding of the absolute significance of smallness is the basic clue to the understanding of quantum mechanics.

In the physics of classical systems—systems to which classical concepts can be applied successfully—it is tacitly assumed that the operations of observation do not appreciably disturb their motion. In applying Maxwell's equations for instance, it is assumed that the currents and fields involved can all be measured without altering their values, or upsetting the development of the observed system. More precisely, it is assumed that any disturbances which are caused by measurement (the variation in current, for example, due to applying a voltmeter) can be corrected for exactly, at least in principle

The simplest type of observation is to look at something. This involves shining light on it which, as we have seen, means striking it with photons. If the position is to be determined accurately, the wavelength of the corresponding waves must be sufficiently short, their frequency correspondingly high, and the momentum of the photons consequently above a certain limit. A blow with such a photon may appreciably disturb the observed system if it is small enough. It is conceivable that these disturbances could also be allowed for, but if not, we have immediately an absolute meaning to size. This idea has been expressed precisely by Dirac,† who postulated in general that *"there is a limit to the fineness of our powers of observation and the smallness of the accompanying disturbance—a limit which is inherent in the nature of things, and can never be surpassed by improved technique"*. If the system is large enough for these unavoidable disturb-

† P. A. M. Dirac, *The Principles of Quantum Mechanics*. Oxford University Press, 4th Edition, 1958, p. 3.

ances to be negligible, the tacit assumption of classical physics applies, and the system may be expected to obey classical laws. If, on the other hand, the system is such that these disturbances are appreciable, it is "small" in an absolute sense, and a new theory is required for dealing with it.

The quantum physicist thus appears, if not as a bull in a china shop, at least as a man with his eyes shut, liable to knock down anything he touches, trying to obtain a clear picture of the delicate objects which surround him. Our problem is to set up a physical theory of information collected in this clumsy manner. The surprising thing is that it can be done at all, not that it takes a form which is fundamentally different from classical theory.

The first point is that since the operations of observation affect the physical systems, they may be expected to appear explicitly in the theory. These operations have two main properties:

(i) To each type of observation (e.g. observation of energy, momentum or position) there belongs a set of numbers—the possible results of the observation. We know already, from the energy levels of hydrogen, that these numbers may run over a continuous range, as in classical theory, or take on a set of discrete values.

(ii) Suppose we have two types of observation $\mathscr{A}$ and $\mathscr{B}$. (For example, $\mathscr{A}$ might mean observation of position and $\mathscr{B}$ that of momentum.) We denote observation $\mathscr{B}$ followed by observation $\mathscr{A}$, by $\mathscr{A}\mathscr{B}$. Then $\mathscr{B}\mathscr{A}$ denotes the same types of observation carried out in the opposite order. Since each observation may disturb, and hence affect, the result of the other, the two procedures may well yield different results. We write this symbolically as

$$\mathscr{A}\mathscr{B} - \mathscr{B}\mathscr{A} \neq 0.$$

The value of this expression must be related to the magnitude of the unavoidable disturbances. It is at this point, and with this interpretation, that we expect some new constant to enter the theory, to give a quantitative rather than a qualitative meaning to our absolute definition of smallness. From our experience with Old Quantum Theory it is an obvious conjecture that this new constant will turn out to be Planck's constant, $\hbar$.

§ 3.2 Operators and Observations: Interpretive Postulates

The reader will hardly have failed to notice that the physical properties of observations correspond exactly to the mathematical properties of operators developed in Chapter 2. To each belongs a set of numbers, and the effect of any pair may depend on the order in which they are applied. We thus make the general assumption that the observations $\mathcal{A}$ are represented by operators, $\hat{A}$, there being one operator for each observable property—the energy, the position, etc. The functions on which the operators operate represent the state of the system, and are known as state functions (or wave functions). If the state function is an eigenfunction it is referred to as an eigenstate.

More precisely we make the following interpretive assumptions (to be discussed below):

I(i): *The possible results of an observation $\hat{A}$ are the corresponding eigenvalues a_n.*

I(ii): *An observation $\hat{A}$ on a system in an eigenstate $u_n(x)$ certainly leads to the result a_n.*

I(iii): *The average value of repeated observations $\hat{A}$ on a set of systems, each one in an arbitrary state $\psi(x)$, is*

$$\bar{a}_\psi = \frac{\displaystyle\int_{-\infty}^{\infty} \psi^*(x)\, \hat{A}(x, \partial/\partial x)\psi(x)\, dx}{\displaystyle\int_{-\infty}^{\infty} \psi^*(x)\, \psi(x)\, dx}. \tag{3.1}$$

($\psi^*(x)$ is the complex conjugate of the function $\psi(x)$.)

The first of these assumptions is almost inevitable, since we must clearly identify the numbers associated with the operation $\mathcal{A}$, namely, the results of the observation, with the numbers associated with the corresponding operator $\hat{A}$. The assumption I(ii) is also very plausible since there is a close parallel between the structure of the eigenvalue equation (2.7) and the ideal physical observation, which is an operation, $\hat{A}$, on the system u_n, which leaves it unchanged apart from the production of a number—the result of the observation, a_n.

The postulate I(iii) deals with a more difficult situation. If the system is in a general state $\psi(x)$, an observation $\hat{A}$ must, by I(i), have as its result one of the eigenvalues of $\hat{A}$. Repeated observations $\hat{A}$ on a set of systems, each in the state $\psi(x)$, will produce a statistical distribution of the different eigenvalues, and I(iii) asserts what the average value of this distribution will be. The average must be some number constructed from the operator $\hat{A}$ and the state $\psi(x)$. It must further be consistent with I(ii). In the special case

$$\psi(x) = u_n(x), \tag{3.2}$$

using (2.7), we have†

$$\bar{a}_{u_n} = \frac{\int u_n^*(x)\, \hat{A}\,(x, \partial/\partial x)\, u_n(x)\, dx}{\int u_n^*(x)\, u_n(x)\, dx} \tag{3.2}$$

$$= \frac{\int u_n^*(x)\, a_n\, u_n(x)\, dx}{\int u_n^*(x)\, u_n(x)\, dx} = a_n. \tag{3.3}$$

This is what is required, since for this case repeated observations will always give the same result a_n. The expression given in I(iii) is, in fact, the simplest which satisfies this consistency requirement.

§ 3.3 Physical Postulates

The interpretive postulates given above, set up the machinery for a mathematical representation of quantum observations (accompanied by inevitable disturbances). We now come to two postulates with more direct physical content.

(a) *The Correspondence Principle*

It is clear that there is one condition that quantum mechanics must satisfy. In the limit of the observed systems becoming large and the disturbances becoming negligible, it must go over into classical mechanics. To ensure this we make the first physical postulate.

P(i): *The essentially definitive relations between physical variables in classical mechanics, which do not involve derivatives, are also satisfied by the corresponding quantum operators.*

Thus if $\hat{x}$ and $\hat{p}$ are the position and momentum operators then the operator for the z-component of angular momentum, for example, is

$$\hat{l}_z = \hat{x}\hat{p}_y - \hat{y}\hat{p}_x. \tag{3.4}$$

For a particle, of definite mass, in a classical potential $V(x)$, the energy operator (Hamiltonian) in terms of the position and momentum operators is the sum of the kinetic and potential energy terms,

$$\hat{H} = \frac{\hat{p}^2}{2m} + V(\hat{x}). \tag{3.5}$$

In particular, the Hamiltonian for a quantum harmonic oscillator of angular frequency ω is

$$\hat{H} = \frac{\hat{p}^2}{2m} + \frac{m}{2}\, \omega^2\, \hat{x}^2. \tag{3.6}$$

---

† When not stated explicitly the range of integration is the physical range of the variable of integration.

3

(b) *The Complementarity Principle*

We must now make more precise the general considerations at the
end of § 3.1. If $\hat{A}$ and $\hat{B}$ represent the observation of particular
observables, the inequality (2.14) means physically that there may
be a mutual disturbance between the two observations. This must be
replaced by some equality for particular observables.

The notion of a photon in Old Quantum Theory and the argument
given in § 3.1 suggest that there is a direct connection between the
mutual disturbances of position and momentum measurements, and
that these should be related to $\hbar$. We thus postulate

$$[\hat{x}, \hat{p}] = \alpha\hbar, \tag{3.7}$$

where α is a number to be determined. The simplest, but not the only,
representation of the operator $\hat{x}$ is an ordinary algebraic variable. We
thus replace $\hat{x}$ by x,

$$\hat{x} \to x. \tag{3.8a}$$

According to (2.17), (multiplied by $\alpha\hbar$),

$$\left[x, -\alpha\hbar\frac{\partial}{\partial x}\right] = \alpha\hbar. \tag{3.9}$$

Thus if $\hat{x}$ is represented by (3.8a), (3.7) suggests that

$$\hat{p} \to -\alpha\hbar\frac{\partial}{\partial x}. \tag{3.8b}$$

The eigenvalue equation, (2.7), for the momentum operator with
eigenvalue p is then

$$\hat{p}u_p(x) = \left(-\alpha\hbar\frac{\partial}{\partial x}\right)u_p(x) = pu_p(x). \tag{3.10}$$

The eigenfunctions are thus

$$u_p = \exp\left[-\frac{px}{\alpha\hbar}\right]. \tag{3.11}$$

If we take

$$\alpha = i, \tag{3.12}$$

we then have the space part of a de Broglie wave (1.18) appearing
automatically as the state function of a particle of definite momentum. This is just the type of fundamental relation between particle
and wave which is required. We therefore make the second physical
postulate,

P(ii): $[\hat{x}, \hat{p}] = i\hbar.$ (3.13)

This leads directly, from (3.8a) and (3.8b), to the important representation of these operators.

$$\hat{x} \to x,$$
$$\hat{p} \to -i\hbar \frac{\partial}{\partial x}, \tag{3.14}$$

which is known as the Schrödinger representation.

The more general statement of P(ii) is that any position observable, $\hat{\chi}$, and its corresponding momentum, $\hat{\pi}$, are said to be complementary, and the operators are assumed to satisfy a commutation relation analogous to (3.13),

$$[\hat{\chi}, \hat{\pi}] = i\hbar.$$

§ 3.4 The Schrödinger Equation and Discrete Energy Levels

If we now put together the complementarity principle, P(ii), the correspondence principle, P(i), and the basic interpretive assumption, I(i) (that the possible values of an observable are given by the eigenvalue equation), we arrive at the equation which determines the possible energy values of any system. Explicitly, the energy operator in the Schrödinger representation is

$$H(\hat{x}, \hat{p}) = \hat{H}\left(x, -i\hbar \frac{\partial}{\partial x}\right). \tag{3.15}$$

The energy eigenvalue equation is

$$\hat{H}\left(x, -i\hbar \frac{\partial}{\partial x}\right) u_E(x) = E u_E(x). \tag{3.16}$$

This is Schrödinger's equation.

For a particle free to move along a line in a potential, $V(x)$, the equation reduces to (see (3.5)),

$$\left(-\frac{\hbar^2}{2m} \frac{\partial^2}{\partial x^2} + V(x)\right) u_E(x) = E u_E(x). \tag{3.17}$$

For the corresponding three-dimensional problem, the equation is

$$\left(-\frac{\hbar^2}{2m} \nabla^2 + V(x, y, z)\right) u_E(x, y, z) = E u_E(x, y, z). \tag{3.18}$$

The equation has to be solved subject to the boundary condition that $u_E(x)$ is finite everywhere (in particular at infinity). A more precise statement of the boundary condition and its physical significance is given in the next section, following (3.38).

A particularly important special case of (3.18) is the equation for the possible energy values of the hydrogen atom. Taking the proton

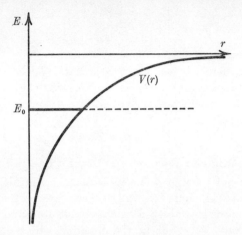

FIG. 3.1. The energy diagram of the Coulomb potential. Physically accessible positions for a classical particle lie above the curve $V(r)$, since points below correspond to negative kinetic energy.

as fixed, we use spherical polar co-ordinates and insert for V the Coulomb potential. Thus

$$\left(-\frac{\hbar^2}{2m}\nabla^2 - \frac{e_M^2}{r}\right)u_E(r, \theta, \phi) = Eu_E(r, \theta, \phi). \qquad (3.19)$$

The crucial test of the theory is that this equation does lead to the observed discrete levels. (This is the exact analogue of the problem of planetary motion for Newtonian mechanics.) The proof of this result involves certain computational difficulties which we postpone to Chapter 7. In the meanwhile we consider a simple model.

The energy diagram for a hydrogen atom is shown in Fig. 3.1. If the kinetic energy is T and the total energy E_0, we have

$$E_0 = T + V.$$

Since, classically, we must have $T \geqslant 0$, the particle of energy E_0 can be found only for those values of r for which the line $E = E_0$ lies above the curve

$$E = V(r).$$

The dotted portion of the line corresponds to negative kinetic energy, and consequently to physically inaccessible positions. For $E_0 > 0$ the electron can reach infinity. For $E_0 < 0$ the electron is bound. For E_0 large and negative the electron is restricted by a very rapidly rising potential to a very narrow region. To investigate the qualitative

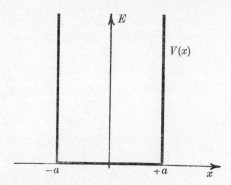

FIG. 3.2. The energy diagram of a one-dimensional infinite square well. A particle is confined to the region $|x| \leqslant a$.

features of such a system we consider the quantum energy values of a particle confined in a one-dimensional infinite square well. Thus

$$V(x) = 0, \; |x| \leqslant a,$$
$$V(x) \to \infty, |x| \geqslant a. \tag{3.21}$$

Classically the particle is confined to the region $|x| \leqslant a$, and, whatever its energy, it bounces elastically off the potential "walls".

The Schrödinger equation for such a system is (3.17) with V defined by (3.21). For $|x| \leqslant a$,

$$-\frac{\hbar^2}{2m}\frac{\partial^2}{\partial x^2}u_n(x) = E_n u_n(x). \tag{3.22}$$

Since V becomes infinite for $|x| \geqslant a$, but other terms in the Schrödinger equation for this region remain finite, we must impose the boundary condition

$$u_n(x) = 0, \qquad |x| \geqslant a. \tag{3.23}$$

Introducing

$$k_n^2 = \frac{2mE_n}{\hbar^2}, \tag{3.24}$$

the equation is

$$\left(\frac{\partial^2}{\partial x^2} + k_n^2\right) u_n(x) = 0. \tag{3.25}$$

The solutions, which satisfy the boundary condition at $x = a$, are

$$u_{2n}(x) = A \sin k_{2n} x, \tag{3.26}$$

where

$$ak_{2n} = 2n(\pi/2), \qquad n = 1, 2, \ldots; \tag{3.27}$$

and

$$u_{2n+1}(x) = B \cos k_{2n+1} x \tag{3.28}$$

where

$$ak_{2n+1} = (2n+1)(\pi/2), \qquad n = 0, 1, 2, \ldots \tag{3.29}$$

Combining (3.27) and (3.29),

$$k_n = \left(\frac{\pi}{2a}\right) n, \qquad n = 1, 2, \ldots$$

and, by (3.24), the possible energy levels are

$$E_n = \frac{\hbar^2}{2m}\left(\frac{\pi^2}{4a^2}\right) n^2. \tag{3.30}$$

The essential feature of a discrete energy spectrum has appeared naturally from the formalism. This may be compared with the Bohr formula for hydrogen,

$$E_n^{(H)} = -\frac{\hbar^2}{2m}\frac{1}{a_0^2}\frac{1}{n^2}. \tag{3.31}$$

Considering the crudeness of the model the similarity is striking. The discrepancy by factors of π is typical of one-dimensional approximations to three-dimensional systems. The difference in sign is due to the fact that for hydrogen the levels are measured from the top of the potential well downwards. For the square well, we measure from the bottom up.

We have now achieved the three major objectives specified at the end of Chapter 1. We have a fundamental reformulation of the theory of atomic systems, which is physically based on Planck's notion of photons and the implication of minimum disturbances accompanying observations. The de Broglie waves have appeared as the eigenfunctions of particles of definite momentum. We have not explicitly produced the Bohr levels, but a direct application of the theory to the crude model of an infinite square well has produced the main qualitative features, the most important being, of course, a discrete energy spectrum.

§ 3.5 The State Functions and Overlap Integral

We now return to a consideration of the state functions and their physical significance. The first remark is that all the physical properties stated so far (I(i), (ii) and (iii)) are unaltered if a given state function is multiplied by any constant. In order to remove this arbitrariness, it is convenient to impose the normalizing condition,

$$\int \psi^*(x)\,\psi(x)\,dx = 1, \tag{3.32}$$

where the integral is taken over all allowed values of x, and here, as elsewhere, * denotes complex conjugation. Thus the state function of a particle of momentum p is the de Broglie wave,

$$u_p(x) = C\,e^{ipx/\hbar}.$$

If the particle is confined to a region

$$0 < x < L,$$

the normalizing condition is

$$\int\limits_0^L |C\,e^{ipx/\hbar}|^2\,dx = 1.$$

Therefore

$$|C|^2 = L^{-1},$$

and the normalized state function is

$$u_p(x) = \frac{1}{L^{1/2}}e^{ipx/\hbar}. \tag{3.33}$$

For normalized state functions the expression (3.1) for the average value of repeated observations, $\hat{A}$, takes on the slightly simpler form

$$\bar{a}_\psi = \int \psi^*(x)\,\hat{A}\left(x, \frac{\partial}{\partial x}\right)\psi(x)\,dx. \tag{3.34}$$

Thus, for example, the average value of the momentum is

$$p_\psi = \int \psi^*(x)\left(-i\hbar\frac{\partial}{\partial x}\right)\psi(x)\,dx. \tag{3.35}$$

Applying this formula to the special case

$$\hat{A} = \hat{x}, \tag{3.36}$$

we get for the average value of the position,

$$\bar{x}_\psi = \int x |\psi(x)|^2 \, dx. \tag{3.37}$$

Since $|\psi(x)|^2$ is the weighting factor appropriate to x in the calculation of the average, it follows that, in a single measurement of the position for a particle in the state $\psi(x)$, the probability of a particular result x is

$$\mathscr{P}_\psi(x) = |\psi(x)|^2. \tag{3.38}$$

Thus the most direct physical interpretation of the state function is that its modulus square determines the probability density of the particle in space. The normalizing condition ensures that the total probability of finding the particle somewhere is unity, and a more general statement of the boundary condition to be placed on any state function is that it can be normalized.

Note that the relative probability for finding the particle at two different positions does not depend on the normalization constant, since it will cancel when the ratio of the two probabilities is taken. In many physical situations, for example the discussion of the uncertainty principle in the next section, only relative probabilities are important, and the normalization constant is irrelevant.

One can reasonably ask a similar question for other observables. Given a general state function $\psi(x)$, what is the probability $\mathscr{P}_\psi(a_n)$ that a particular observation $\hat{A}(x, \partial/\partial x)$ will have the result a_n? The answer to this can, in fact, be deduced from the postulates already made, as is shown below in Chapter 12. We content ourselves here with stating the result, and making it physically plausible. If the state function happens to be the eigenstate corresponding to a_n,

$$\psi(x) = u_{a_n}(x),$$

then the probability is unity. In general the probability must be related to the extent to which the state function $\psi(x)$ resembles the eigenfunction $u_{a_n}(x)$. This is given quantitatively by the *overlap integral*,

$$\int u_{a_n}^*(x) \, \psi(x) \, dx.$$

This is a number, which is unity in the special case quoted above; is **near unity** when the two functions are similar (large overlap), and

very small when the overlap is small (see Fig. 3.3). The probability we
are seeking must be real and this is ensured if we take the modulus
square. The required probability is

$$\mathscr{P}_\psi(a_n) = |\int u_{a_n}^*(x)\,\psi(x)\,dx|^2. \tag{3.39}$$

The state functions, $\psi(x)$, contain all the information about a
system which it is possible to have, consistent with the mutual
disturbances of observations. This information is partly statistical in
character because of the random effects of these disturbances. While

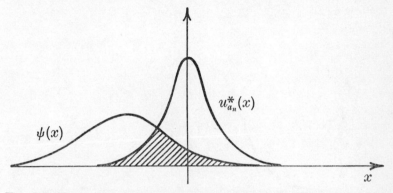

Fig. 3.3. The overlap integral of $u_n^*(x)$ and $\psi(x)$ comes from the shaded
region in which neither factor is zero.

unobserved (and hence undisturbed) the systems are assumed to
develop in a causal manner, according to differential equations which
are set up in Chapter 13.

It is usually adequate to think of the state functions as describing
the actual physical systems. Occasionally it is important to remember
that these functions describe, more precisely, an ideal state of knowledge of a system. Thus if, for example, an energy observation is made
on a system in a state $\psi(x)$, the result must be some eigenvalue E_n.
The making of the measurement produces an abrupt change in our
state of knowledge of the system, and it is subsequently described by
the corresponding eigenstate $u_{E_n}(x)$.

§ 3.6 The Uncertainty Principle

The last point to be discussed in this fundamental chapter is the
statement made at the end of § 3.3, that the commutation relation,
(3.13), for $[\hat{x},\hat{p}]$ is an assertion about the mutual disturbance between
these two types of observation.

To do this, consider the state function

$$\psi(x) = \exp[-x^2/2\Delta_x^2]. \tag{3.40}$$

(This is not normalized and the normalization does not effect the argument.) Then the relative probability density is, by (3.38),

$$\mathscr{P}_\psi(x) = |\psi(x)|^2 = \exp[-x^2/\Delta_x^2], \tag{3.41}$$

which is a Gaussian hump of half width Δ_x. Thus $\psi(x)$ represents a particle which is almost certainly situated within a distance Δ_x of the origin.

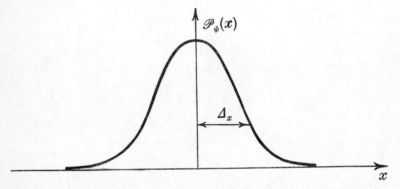

FIG. 3.4. The probability distribution corresponding to the state function (3.40). The particle is almost certain to be found within a distance Δ_x of the origin.

If a measurement is made of the momentum, the probability of a result p is given in terms of the overlap integral (3.39), which we call $\phi(p)$. Thus

$$\phi(p) = \int\limits_{-\infty}^{\infty} u_p^*(x)\,\psi(x)\,dx,$$

$$= \int\limits_{-\infty}^{\infty} e^{-ipx/\hbar}\,\psi(x)\,dx, \tag{3.42}$$

and the probability of momentum p is,

$$\mathscr{P}_\psi(p) = |\phi(p)|^2. \tag{3.43}$$

Because of the analogy between (3.43) and (3.38), $\phi(p)$ is called the state function in momentum space. Substituting for $\psi(x)$ in (3.42),

$$\phi(p) = \int\limits_{-\infty}^{\infty} \exp\left[-\frac{ipx}{\hbar}\right]\exp\left[-\frac{x^2}{2\Delta_x^2}\right]dx$$

$$= \int\limits_{-\infty}^{\infty} \exp\left[-\frac{1}{2}\left(\frac{x}{\Delta_x}+\frac{ip\Delta_x}{\hbar}\right)^2\right]dx\exp\left[-\frac{p^2\Delta_x^2}{2\hbar^2}\right].$$

Putting

$$\frac{x}{\Delta_x}+\frac{ip\Delta_x}{\hbar} = y, \tag{3.44}$$

the integral becomes $\int\limits_{-\infty}^{\infty} \exp[-\frac{1}{2}y^2]dy$, which is simply a constant that can be absorbed in the normalization. Hence, $\phi(p)$, with arbitrary normalization, is

$$\phi(p) = \exp[-p^2\Delta_x^2/2\hbar^2]. \tag{3.45}$$

If we define Δ_p by the relation

$$\Delta_x\Delta_p = \hbar, \tag{3.46}$$

then

$$\phi(p) = \exp[-p^2/2\Delta_p^2], \tag{3.47}$$

showing, in precise analogy with $\psi(x)$, that the particle almost certainly has a momentum which differs from zero by at most Δ_p.

The relation (3.46) between the uncertainty in the position, Δ_x, and the uncertainty in the momentum, Δ_p, is a direct consequence of the commutation relation between $\hat{x}$ and $\hat{p}$, (3.13). The precise equality (3.46) depends on the Gaussian form, which we chose for the state function to simplify the algebra. The general statement, which applies to any state function, which specifies the position of the particle within approximate limits Δ_x, and the momentum within Δ_p, is

$$\Delta_x\Delta_p \geqslant \hbar. \tag{3.48}$$

This is known as the *uncertainty principle*. This is a more precise statement of the magnitude of mutual disturbances between complementary variables. The smaller Δ_x, the more accurately the position is known, the larger the disturbance to p and hence the larger the uncertainty Δ_p. The minimum value for the product of the two uncertainties is determined by Planck's constants.

We already have an example of one extreme case of this. The state function (de Broglie wave) for a particle of momentum p, normalized to lie somewhere on a line of (large) length L, is given by (3.33). Since p is known exactly for this state,

$$\Delta_p = 0. \tag{3.49}$$

The position probability density is

$$\mathscr{P}_{u_p}(x) = |u_p(x)|^2 = \frac{1}{L}. \tag{3.50}$$

Since this is independent of x, the particle is equally likely to be anywhere. In the limit $L \to \infty$, we have

$$\Delta_x \to \infty, \tag{3.51}$$

in accordance with the uncertainty principle.

A great deal of attention is usually paid to the uncertainty principle, but it is an essentially negative statement, expressing the limitations on our knowledge imposed by the mutual disturbances of observations. In general, if we have two observations $\hat{A}$ and $\hat{B}$ and

$$[\hat{A}, \hat{B}] \neq 0, \tag{3.52}$$

then these disturbances prevent one from ever having an exact knowledge of the result of both types of observations simultaneously. Much more important in the long run is the corresponding positive statement that if

$$[\hat{A}, \hat{B}] = 0, \tag{3.53}$$

there is no mutual disturbance, and the result of both types of observation can be known exactly.

An example of this is given by the momentum and energy for a free particle

$$\hat{A} = \hat{p} \to -i\hbar \frac{\partial}{\partial x}, \tag{3.54}$$

$$\hat{B} = \hat{H} = \frac{\hat{p}^2}{2m} \to -\frac{\hbar^2}{2m} \frac{\partial^2}{\partial x^2}. \tag{3.55}$$

Then

$$[\hat{p}, \hat{H}] = \left[-i\hbar \frac{\partial}{\partial x}, -\frac{\hbar^2}{2m} \frac{\partial^2}{\partial x^2} \right]$$

$$\equiv \frac{i\hbar^3}{2m} \left(\frac{\partial}{\partial x} \frac{\partial^2}{\partial x^2} - \frac{\partial^2}{\partial x^2} \frac{\partial}{\partial x} \right) = 0, \tag{3.56}$$

which is an operator equation, meaning that the differential operator in the final brackets gives zero, when operating on *any* function $\psi(x)$. Thus the energy and momentum of a free particle can be known exactly, simultaneously. To establish this formally, we know that the eigenstate of momentum p is the de Broglie wave

$$u_p = e^{ipx/\hbar}. \tag{3.57}$$

Now

$$\hat{H}u_p = -\frac{\hbar^2}{2m}\frac{\partial^2}{\partial x^2}e^{ipx/\hbar} = \frac{p^2}{2m}u_p = Eu_p, \tag{3.58}$$

which is the eigenvalue equation for energy. Therefore u_p is also an eigenstate of energy, corresponding to the eigenvalue

$$E = p^2/2m. \tag{3.59}$$

Note that this it not true if the particle moves in a potential $V(x)$. In this case

$$[\hat{H}, \hat{p}] \neq 0, \tag{3.60}$$

since

$$[V(\hat{x}), \hat{p}] \neq 0,$$

so that, if the energy of the particle is known exactly, it will not have definite momentum, only an average value defined by I(iii)—and vice versa.

If the system is large so that $\hbar$ may be neglected, all the operators commute with each other, so that all observables can be measured without mutual disturbance. This finds its mathematical expression in the fact that all operators may then be represented by ordinary algebraic variables. The correspondence principle ensures that the definitive relations between these variables—energy, momentum, etc.—are the same as in classical theory, so we go over smoothly into the classical formalism. (It is shown in Chapter 13 that this is also true for relations involving time derivatives.)

One final point that should perhaps be emphasized, is that in all the above discussion the distinction between "small" systems and "large" systems—quantum systems and classical systems—is not made on the basis of spatial extension only, but in units of $\hbar$. This has the dimensions of "action", (ML^2T^{-1}), which is

$$(\text{length}) \times (\text{momentum}),$$

or

$$(\text{time}) \times (\text{energy}),$$

and it is in terms of the typical action that the "size" of a system must be judged. Thus, for example, for an electron in an atom the typical "action" is the product of the Bohr radius, $\sim 10^{-10}$ m, and the momentum in the Bohr orbit, 10^{-24} mks (see Problem **1.2**). This is just of the order of $\hbar$, so quantum mechanics is essential. On the other hand, in most situations in electronics the distances involved are of the order of 10^{-4} m. Voltages are typically tens of volts, and momenta are consequently of 10 eV/c $\sim 10^{-26}$ mks, giving an action of $\sim 10^{-30}$ mks, which is safely in the classical region. Thus, such electrons may be treated as classical particles. There are, however, exceptional situations in electronics in which $\hbar$ cannot be neglected, and quantum effects are important, as in the tunnel diode, which employs the purely quantum phenomenon of barrier penetration, discussed towards the end of §. 4.1.

§ 3.7 Summary

This completes our setting up of quantum mechanics. The previous sections contain a great number of new concepts and are bewildering at first sight. The only way to get over this is to use the new formalism, and make oneself familiar with it. Familiarity does not breed contempt, but wonder at the beautiful way in which all the pieces of the complicated puzzle fit together.

As stated at the beginning of this chapter, we have developed a theory of a completely new type of information. The tacit assumption of classical physics—that things can be observed without disturbing them—is also the tacit assumption of everyday life, particularly of visual information. Our minds are accustomed to deal with information gathered in this way. It is the intellectual leap which is required to train ourselves to handle this new type of "quantum" information, which is the main difficulty in gaining an initial understanding of quantum mechanics.

In conclusion we summarize very briefly the steps in the argument:

(i) The notion of *photons* implies unavoidable disturbances accompanying observations. Quantum mechanical systems are those which are appreciably disturbed by observation.

(ii) The operations of observation must appear explicitly in the theory. Classical observables such as energy, H, momentum, p, and position, x, are replaced by quantum operators, $\hat{H}, \hat{p}, \hat{x}$. The definitive relations between them are the same in classical and quantum mechanics (correspondence principle, P(i)).

(iii) The possible values of an observable $\hat{A}$ are the eigenvalues a given by the eigenvalue equation,

$$\hat{A}u_a(x) = au_a(x);$$

measurement on an eigenstate $u_a(x)$ leading certainly to a result a (I(i) and I(ii)). The equation is solved subject to the boundary condition that the state function can be normalized,

$$\int |u(x)|^2 dx = 1.$$

(iv) The average value of repeated observations $\hat{A}$ on systems in an arbitrary state $\psi(x)$ (normalized) is

$$\bar{a}_\psi = \int \psi^*(x)\, \hat{A}\left(x, \frac{\partial}{\partial x}\right) \psi(x)\, dx. \qquad \text{I(iii)}$$

Applying this to the observation of position, implies that the probability of position x is

$$\mathscr{P}_\psi(x) = |\psi(x)|^2.$$

Normalization of the state function ensures that the total probability of finding the particle somewhere is unity.

(v) The probability of an observation $\hat{A}$ on a state $\psi(x)$ having a result a is

$$\mathscr{P}_\psi(a) = \left| \int u_a^*(x)\, \psi(x)\, dx \right|^2,$$

the integral being the *overlap integral*.

(vi) The mutual disturbance between observations of the complementary variables, position and momentum, is measured by Planck's constant, $\hbar$, and is expressed formally by

$$[\hat{x}, \hat{p}] = i\hbar. \qquad \text{P(ii)}$$

This leads directly to the Schrödinger representation of these operators

$$\hat{x} \to x,$$

$$\hat{p} \to -i\hbar \frac{\partial}{\partial x}.$$

This in turn shows that the eigenstate, corresponding to a particle of definite momentum p is the *de Broglie Wave*,

$$u_p(x) = e^{ipx/\hbar}.$$

(vii) Substituting the Schrödinger representation for $\hat{x}$ and $\hat{p}$ into the eigenvalue equation for the energy, $H(\hat{x}, \hat{p})$, then gives the *Schrödinger equation* for the possible energy levels of a system.

$$\hat{H}\left(x, -i\hbar \frac{\partial}{\partial x}\right) u_E(x) = E u_E(x).$$

When applied to the hydrogen atom this correctly predicts the *Bohr levels*.

(viii) The commutation relation for $[\hat{x}, \hat{p}]$ in (vi), implies the *uncertainty relation* between position and momentum,

$$\Delta_x \Delta_p \geqslant \hbar.$$

PROBLEMS III

3.1. A particle confined in an infinite square well potential (3.21) has an uncertainty in its position

$$\Delta_x = 2a.$$

The magnitude of the momentum must be at least as large as the uncertainty in the momentum. Hence, show that an estimate of the ground state energy, based on the uncertainty principle is

$$E_1 \simeq \frac{\hbar^2}{8ma^2}.$$

Compare this with the exact value obtained from the eigenvalue equation.

3.2. By using the Schrödinger representation of the operators (3.14), show that

$$[\hat{p}, V(\hat{x})] = -i\hbar \frac{\partial V(\hat{x})}{\partial \hat{x}}.$$

(This is a direct generalization of the argument that led to (2.17).)

3.3. Given that

$$\int\limits_{-\infty}^{+\infty} e^{-x^2 a}\, dx = \sqrt{\left(\frac{\pi}{a}\right)},$$

derive (3.45) from (3.40) including the normalization constants.

3.4. A state function (not normalized) of a particle free to move on a straight line is

$$\psi(x) = \exp\left[\frac{-x^2}{2\Delta^2} + i\frac{Px}{\hbar}\right].$$

Show that the average value of the momentum is P, and that the uncertainty in the position of the particle is of order Δ. By considering the state function in momentum space, show that the momentum is unlikely to differ from P by an amount greater than $\hbar/\Delta$.

3.5. A particle of mass m is confined by an infinite square well potential as defined in (3.21). If the particle is in the state

$$\psi(x) = x, \qquad |x| \leqslant a,$$

$$\psi(x) = 0, \qquad |x| > a,$$

find the relative probability that a measurement of the energy will give the result E_2 or E_4.

3.6. A quantum particle, free to move on a straight line, has state function

$$\psi(x) = \left(\frac{1}{2a}\right)^{1/2}, \qquad |x| \leqslant a,$$

$$\psi(x) = 0, \qquad |x| > a;$$

find the probability (arbitrarily normalized) for the particle to be found with momentum p. Make a rough sketch of this momentum probability distribution and discuss it in relation to the corresponding spatial distribution and the uncertainty principle.

Show that the relative probability of finding the particle with momentum $\pi\hbar/2a$ and zero is $4/\pi^2$.

CHAPTER 4

ONE-DIMENSIONAL MOTION

§ 4.1 The Potential Step

Before considering the quantum theory of the harmonic oscillator and hydrogen atom, we compare the quantum and classical predictions in some simple one-dimensional cases.

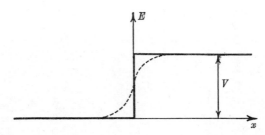

FIG. 4.1. The energy diagram of a potential step. The dotted curve gives the realistic situation. The full line is an idealized situation, for which calculations are easier.

The simplest system is the motion of a particle in a potential of the form illustrated in Fig. 4.1 by the dotted curve. Since the force $F(x)$ is

$$F(x) = -\frac{\partial V}{\partial x}, \qquad (4.1)$$

this represents a particle which moves freely except in the neighbourhood of the origin, where it is subjected to a force towards the left. If the particle has total energy E_0, and kinetic energy T, then

$$E_0 = T(x) + V(x). \qquad (4.2)$$

There are two cases to be considered and we look at them both classically first.

(i) $E_0 > V$ (Classical)

Particles coming from the left approach the potential barrier with kinetic energy T_0, and momentum p_0, given by

$$T_0 = E_0 = \frac{p_0^2}{2m}.$$

36

As the particles move through the region of the potential barrier, they are slowed up by the force, and kinetic energy is converted into potential energy. They have sufficient energy to penetrate the barrier and there is *total transmission*. The particles emerge to the right with kinetic energy, T_1, and momentum, p_1, where

$$T_1 = \frac{p_1^2}{2m} = E_0 - V. \tag{4.3}$$

(ii) $E_0 < V$ *(Classical)*

Particles coming from the left are stopped by the potential barrier at the point x' where

$$V(x') = E_0, \qquad [T(x') = 0]. \tag{4.4}$$

Their motion is then exactly reversed under the action of the force. In this case there is, therefore, *total reflection* of the beam.

The qualitative features of the classical motion are unchanged if the potential barrier is replaced by a sudden potential step, as illustrated by the solid line in Fig. 4.1. This is a simpler system to discuss quantum mechanically.

The quantum mechanical motion is determined by the energy eigenvalue equation

$$\left(\frac{\hat{p}^2}{2m} + V(\hat{x})\right) u_E = E_0 u_E \tag{4.5}$$

where

$$V(x) = 0, \qquad x < 0,$$

$$V(x) = V, \qquad x > 0. \tag{4.6}$$

In the Schrödinger representation this is

$$\left(-\frac{\hbar^2}{2m}\frac{\partial^2}{\partial x^2} + V(x)\right) u_E(x) = E_0 u_E(x). \tag{4.7}$$

We have the general boundary condition that $u_E(x)$ must be finite everywhere. We have also to consider the discontinuity at $x = 0$. Since the sudden change in $V(x)$ is finite, and $u(0)$ is finite, the equation (4.7) asserts that $u''(0)$ (second derivative) is finite. This implies that both $u(x)$ and $u'(x)$ are continuous at $x = 0$. We must now distinguish the two cases.

Case (i) $E_0 > V$ *(Quantal)*

Define

$$k_0^2 = \frac{2mE_0}{\hbar^2}, \tag{4.8}$$

and

$$k_1^2 = \frac{2m(E_0 - V)}{\hbar^2}. \tag{4.9}$$

Then equation (4.7) becomes

$$\left(\frac{\partial^2}{\partial x^2} + k_0^2\right) u_L(x) = 0, \qquad x < 0; \tag{4.10}$$

and

$$\left(\frac{\partial^2}{\partial x^2} + k_1^2\right) u_R(x) = 0, \qquad x > 0; \tag{4.11}$$

the suffixes L and R denoting the solutions to left and right of the origin, respectively. The solutions are linear combinations of

$$u_L(x) = e^{\pm i k_0 x}, \tag{4.12}$$

$$u_R(x) = e^{\pm i k_1 x}. \tag{4.13}$$

These are de Broglie waves corresponding to the momenta p_0 and p_1 of the classical problem. We are interested in the situation in which a particle approaches from the left, and may then be either transmitted or reflected. We thus look for a solution of the form

$$u_L(x) = \quad e^{i k_0 x} \quad + A\, e^{-i k_0 x}, \tag{4.14}$$

$$\text{(incident)} + \text{(reflected)}.$$

$$u_R(x) = \quad B\, e^{i k_1 x}, \tag{4.15}$$

$$\text{(transmitted)}.$$

We have arbitrarily normalized the coefficient of the incident wave to unity. We want to find the possible energy values, E_0, of the system, and the reflected and transmitted intensities, determined through A and B, respectively.

The continuity conditions at $x = 0$ are

$$1 + A = B, \quad (u \text{ continuous}), \tag{4.16}$$

$$k_0(1 - A) = k_1 B, \qquad (u' \text{ continuous}). \tag{4.17}$$

These can be solved for any value of E_0, and imply

$$A = \frac{k_0 - k_1}{k_0 + k_1}, \qquad B = \frac{2k_0}{k_0 + k_1}. \tag{4.18}$$

Thus, provided $E_0 > V$, the system can have any energy, as in the classical theory. However, the possible motion differs from the classical in an essential way.

The relative probability of finding the particle at a point x, for $x < 0$, is

$$\mathcal{P}_\psi(x) = |e^{ik_0 x} + A e^{-ik_0 x}|^2$$
$$= 1 + |A|^2 + 2A \cos 2k_0 x. \tag{4.19}$$

The final oscillating term is not of much physical interest, and can be removed by averaging over a region of length large compared with $2\pi/k_0$. The other two terms come directly from the incident and reflected beams, and can be interpreted as the relative intensities of these beams. By the same token $|B|^2$ is the relative intensity of the transmitted beam. The important new qualitative feature of the quantum theory is that, since

$$|A|^2 \neq 0, \tag{4.20}$$

there is a non-vanishing reflected beam.

Two limiting cases are of particular interest. The condition for quantum mechanics to be necessary is

$$\text{(typical length)} \times \text{(typical momentum)} \leqslant \hbar.$$

The typical length is the distance through which the potential is changing; the typical momentum is that of the incoming beam. Thus the classical limit is obtained for large momenta, or

$$E_0 \gg V. \tag{4.21}$$

In this case, by (4.8) and (4.9),

$$k_0 \simeq k_1.$$

Hence, by (4.18),

$$A \simeq 0, \qquad B \simeq 1, \tag{4.22}$$

which is the correct classical limit of total transmission.

The extreme quantum limit is

$$E_0 \ll |V|. \tag{4.23}$$

The most interesting case is to take V large in magnitude, but negative, so that we have a sudden large potential drop, through which classical particles pass with greatly increased momentum. However, quantum mechanically, by (4.8) and (4.9), we now have

$$k_0 \ll k_1. \qquad (4.24)$$

Thus, by (4.18),

$$A \simeq -1, \qquad B \simeq 0, \qquad (4.25)$$

implying total reflection—the exact opposite of the classical prediction. This essentially quantum effect may be observed in nuclear physics when a low energy incident neutron, say, is *reflected* by the sudden onset of the highly attractive potential, as it approaches the surface of a nucleus.

Case (ii) $E_0 < V$ *(Quantal)*

The equation for $u_L(x)$, $(x < 0)$, is exactly as before. For $u_R(x)$ we now define

$$K^2 = \frac{2m}{\hbar^2}(V - E_0), \qquad (4.26)$$

and then

$$\left(\frac{\partial^2}{\partial x^2} - K^2\right) u_R(x) = 0, \qquad x > 0. \qquad (4.27)$$

Again, to represent a state with unit intensity incident beam from the left, we have a solution of the form

$$u_L(x) = e^{ik_0 x} + A e^{-ik_0 x}, \qquad (4.28)$$

$$u_R(x) = C e^{-Kx} + D e^{+Kx}. \qquad (4.29)$$

To satisfy the condition that u is normalizable, u_R must go to zero at infinity, so we must have

$$D = 0.$$

The continuity conditions at $x = 0$ are

$$1 + A = C, \qquad (4.30)$$

$$ik_0(1 - A) = -KC, \qquad (4.31)$$

which can be solved to give

$$A = \frac{k_0 - iK}{k_0 + iK}, \qquad C = \frac{2k_0}{k_0 + iK} \qquad (4.32)$$

for any value of E_0. Thus again there is no restriction on the possible energy values. Further

$$|A|^2 = 1, \qquad (4.33)$$

so that for any energy, in this range, we have total reflection, as in the classical case.

However, the relative probability for finding the particle in the classically forbidden region, $x > 0$, is

$$\mathscr{P}_{u_R}(x) = |C e^{-Kx}|^2$$

$$= \frac{4k_0^2}{k_0^2 + K^2} e^{-2Kx}. \tag{4.34}$$

This is appreciable near to the barrier edge, and falls off exponentially to a negligible value over a distance large compared with $1/K$.

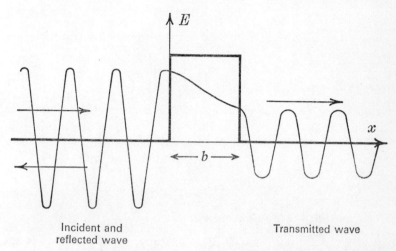

Incident and reflected wave

Transmitted wave

FIG. 4.2. The energy diagram of a finite potential barrier, showing the incident and transmitted wave.

This is particularly important if we consider a wave impinging on a barrier of finite thickness b (Fig. 4.2). This effect will give an appreciable probability for finding the particle at the opposite edge of the potential barrier, and it will then propagate to the right as a free particle. From (4.34) the relative probability of finding the particle at $x = b$ and at $x = 0$ is

$$T = \exp[-2Kb]$$

$$= \exp\left[-2\left(\frac{2m}{\hbar^2}\right)^{1/2} (V - E_0)^{1/2} b\right]. \tag{4.35}$$

This is an approximate expression, valid for large b, for the probability that a particle, of energy E_0, penetrates a barrier of height V and width b. (The solution inside a barrier of finite width is really a mixture of rising and falling exponential, $\exp[\pm Kx]$, and for narrow barriers there is a considerable mixture of rising exponential. (See Problem 4.4.)) This quantum possibility of penetrating potential barriers, which will certainly stop classical particles, is the basis of the explanation of the radioactive decay of a nucleus (Chapter 9).

§ 4.2 Parity

Before considering other potentials, it is convenient to introduce some general considerations of the state functions for systems with potentials such that

$$V(x) = V(-x). \tag{4.36}$$

Such potentials are symmetric about the origin. Given any solution $u_E(x)$,

$$\left(\frac{-\hbar^2}{2m}\frac{\partial^2}{\partial x^2} + V(x)\right)u_E(x) = Eu_E(x), \tag{4.37}$$

by changing x to $-x$, and using (4.36), it is simple to show that for such systems $u_E(-x)$ is also a solution.

If we assume there is only one linearly independent solution for a given eigenvalue E, then the two solutions can differ only by a constant. Therefore

$$u_E(x) = \epsilon u_E(-x). \tag{4.38}$$

Changing x to $-x$ in this equation,

$$u_E(-x) = \epsilon u_E(x). \tag{4.39}$$

Substituting (4.39) into (4.38), shows that

$$\epsilon^2 = 1, \qquad \epsilon = \pm 1. \tag{4.40}$$

The constant ϵ is known as the *parity* of the state. For positive parity states, ($\epsilon = +1$), $u_E(x)$ is symmetric about the origin; for negative parity, ($\epsilon = -1$), $u_E(x)$ is anti-symmetric.

If there is more than one solution, and a particular solution does not have definite parity, then from $u_E(x)$ and $u_E(-x)$ one can always construct solutions

$$u_e(x) = \tfrac{1}{2}[u_E(x) + u_E(-x)],$$
$$u_o(x) = \tfrac{1}{2}[u_E(x) - u_E(-x)], \tag{4.41}$$

each of which has definite parity. Thus it is always possible to construct solutions, which are either symmetric or anti-symmetric.

§ 4.3 Bound States

So far we have considered potentials in which the particles were free, in the sense that both classically and according to quantum mechanics the particle could move off to infinity in at least one direction. We now consider a bound system, in which this is not the case. The simplest to discuss mathematically is again a "square" well potential given by (Fig. 4.3)

$$V(x) = 0, \qquad |x| \leqslant a,$$
$$V(x) = V, \qquad |x| > a. \tag{4.42}$$

We consider the case $E_0 < V$, so that classically the particle moves freely in the region $|x| \leqslant a$, but is confined by the potential to this

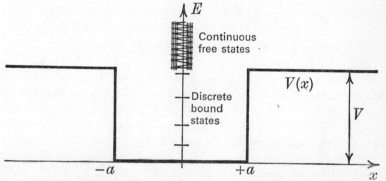

Fig. 4.3. The energy diagram of a square potential well.

region. (We have already considered above, § 3.4, the limiting case of $V \to \infty$.) The potential is symmetric and the knowledge that the eigenfunctions are either symmetric or anti-symmetric is considerable assistance in finding the solutions.

The energy eigenvalue equation is (4.7), which for $|x| \leqslant a$ is identical with (4.10), and for $|x| > a$ takes the form (4.27). If we consider solutions in the "inside" region, $|x| \leqslant a$, we have

$$u_i(x) = \cos k_0 x, \ \sin k_0 x. \tag{4.43}$$

We have deliberately written the even and odd functions, rather than the simple exponentials. In the right-hand outside region

$$u_R(x) = C e^{-Kx}, \tag{4.44}$$

as in (4.29). The alternative solution with a rising exponential is forbidden by the requirement that $u(x)$ goes to zero at infinity, so that it can be normalized. The requirements of parity then show that in the outside left-hand region we must have

$$u_L(x) = \pm C e^{-K|x|}. \tag{4.45}$$

There are thus two possible types of solution, the symmetric ones (with positive parity),

$$u_i(x) = \cos k_0 x, \tag{4.46}$$

$$u_R(x) = C e^{-Kx}, \qquad u_L(x) = C e^{-K|x|};$$

and the anti-symmetric (with negative parity),

$$u_i(x) = \sin k_0 x \tag{4.47}$$

$$u_R(x) = C' e^{-Kx}, \qquad u_L(x) = -C' e^{-K|x|}.$$

In both cases we have arbitrarily normalized the coefficient of the "inside" term to unity. The correct normalization according to (3.32) can always be obtained by multiplying the entire solution by the appropriate constant.

For each solution there are two continuity conditions to be satisfied at $|x| = a$. However, there is only one free constant, because the rising exponential, unlike the alternative oscillation in the free case, has to have zero coefficient. This means that to satisfy the continuity conditions, the energy E_0 must also be regarded as an adjustable parameter. Only for certain discrete values, E_n, can the continuity conditions be satisfied. The equations for the even solution are

$$\cos k_0 a = C e^{-Ka}, \tag{4.48}$$

$$k_0 \sin k_0 a = KC e^{-Ka},$$

which can only be satisfied if

$$k_0 \tan k_0 a = K. \tag{4.49}$$

Substituting from (4.8) and (4.26) for k_0 and K, this equation determines the values of E_n corresponding to even parity solutions. Similarly the odd parity solutions are given by

$$k_0 \cot k_0 a = -K. \tag{4.50}$$

These equations may be solved graphically for E_n. However, the most important feature is not the particular solutions, but the general one that those ranges of energy, which classically imply bound states, lead in quantum mechanics to a discrete energy spectrum.

It is also possible to make some general remarks about the form of the eigenfunctions. According to (4.27), in the outside region

$$u''/u > 0. \tag{4.51}$$

This means that for u positive the curve has the general features of a minimum; for u negative, those of a maximum. These statements can

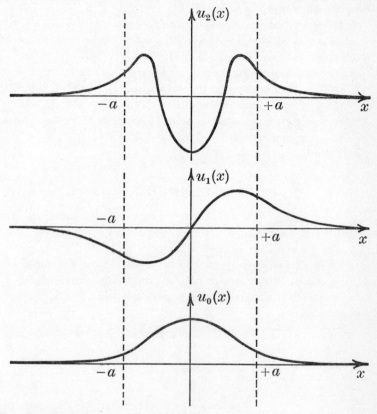

Fig. 4.4. Typical shape of the energy eigenfunctions of the first three bound levels showing parity and number of nodes.

be combined into the single statement that in the outside region the state function curves away from the axis (convex). Similarly in the inside region the state function curves towards the axis (concave). Further, the ratio u''/u determines the curvature, and according to the eigenvalue equation, this increases with E. The eigenvalues E_n are those particular values for which the outside convex parts of the state function match up in slope and value with the inside concave parts.

The ground state $u_0(x)$ is an even state of minimum curvature with no nodes. The next state $u_1(x)$ is an odd state with a single node at the origin. The state $u_2(x)$ is again even with two nodes, as illustrated in Fig. 4.4. In general the nth state has parity $(-1)^n$, with n nodes.

Summarizing, we have found that:

   (i) for energy ranges corresponding classically to free states, the quantum system can have the same continuous range of energies;

 (ii) particles moving in symmetric potentials have energy eigenstates which are symmetric (even parity), or anti-symmetric (odd parity), about the origin;

(iii) for energy ranges corresponding classically to bound states, the quantum system has discrete energy levels;

(iv) the nth level has parity $(-1)^n$, and n nodes.

PROBLEMS IV

4.1. Draw rough graphs of the energy eigenfunctions of the infinite potential well and verify that they satisfy conditions (ii), (iii), and (iv) of the summary to this chapter.

4.2. Show that energy eigensolutions to the symmetric square well potential can be found for any value of the energy provided $E > V$.

4.3. A particle of mass m, free to move on a straight line, approaches a potential barrier,

$$V(x) = 0, \qquad x < 0 \text{ or } x > a,$$
$$V(x) = V, \qquad 0 \leqslant x \leqslant a,$$

from $x = -\infty$. The energy is $E(< V)$ and

$$k^2 = \frac{2mE}{\hbar^2},$$

$$K^2 = \frac{2m(V-E)}{\hbar^2}.$$

Find the ratio of intensity in the transmitted and reflected beams if

$$Ka \ll 1.$$

If the beam approaches the "middle" of a very narrow barrier so that also

$$k \simeq K \quad \text{and} \quad ka \ll 1,$$

show that the beam is almost completely transmitted. On the other hand if the beam approaches the top of a high barrier so that

$$k \gg K \quad \text{and} \quad ka \gg 1,$$

show that the beam is almost completely reflected. (*Hint.* Express the state function for $x > a$ as $u(x) = D e^{ik(x-a)}$ and expand $e^{Ka} \simeq 1 + Ka$, etc.)

4.4. A particle, of mass m, moves in a potential

$$V(x) \to \infty, \qquad x \leqslant 0,$$

$$V(x) = 0, \qquad 0 < x < a,$$

$$V(x) = V(> 0), \qquad a \leqslant x \leqslant b,$$

$$V(x) = 0, \qquad x > b.$$

If $b \to \infty$ show that the allowed values of the energy for $E < V$ satisfy

$$k \cot ka + K = 0,$$

where k and K are defined in Problem **4.3**. If the energy has such a value, but b is finite, show that the relative intensity at $x = b$ and $x = a$ is

$$T = e^{-2K(b-a)}$$

For the region $x > b$, show that there is equal intensity in the beams travelling to left and right. The state thus represents a reservoir of particles trapped by the potential near the origin at an energy which is a possible eigenvalue in the limit of $b \to \infty$. For finite b the particles are continually leaking through the barrier, but also being replaced by a source at infinity. This system is very closely related to a radioactive nucleus (see § 9.3). (*Hint.* The state function for $0 < x < a$ is $\sin kx$, since $u(0) = 0$. In the other regions take

$$u(x) = A e^{-K(x-a)} + B e^{K(x-a)}, \qquad a < x < b,$$

$$u(x) = C e^{-ik(x-b)} + D e^{+ik(x-b)}, \qquad b < x \,)$$

CHAPTER 5

THE HARMONIC OSCILLATOR

§ 5.1 Classical Theory

According to classical theory a harmonic oscillator is a particle, mass m, moving under the action of a force

$$F = -m\omega^2 x. \tag{5.1}$$

The equation of motion is then

$$\frac{d^2 x}{dt^2} + \omega^2 x = 0, \tag{5.2}$$

with solution

$$x = a \cos \omega t, \tag{5.3}$$

which represents an oscillatory motion of angular frequency ω, and amplitude a. The potential is related to the force by

$$F = -\frac{\partial V}{\partial x},$$

so that

$$V(x) = \tfrac{1}{2} m \omega^2 x^2. \tag{5.4}$$

The energy of the oscillation (5.3) is the potential energy when the particle is at an extreme position. Therefore

$$E = \tfrac{1}{2} m a^2 \omega^2. \tag{5.5}$$

§ 5.2 Quantum Theory: The Eigenvalues

We now consider the quantum theory of such a system. Since the classical motion is bound for all values, the entire quantum energy spectrum should consist of discrete values. The energy eigenvalue equation is (3.16) with

$$\hat{H} = \frac{\hat{p}^2}{2m} + \tfrac{1}{2} m \omega^2 \hat{x}^2. \tag{5.6}$$

In the Schrödinger representation this is

$$\left[\frac{-\hbar^2}{2m} \frac{\partial^2}{\partial x^2} + \frac{m \omega^2}{2} x^2 \right] u_n(x) = E_n u_n(x). \tag{5.7}$$

If this is multiplied by $2/\hbar\omega$ we get

$$\left[\frac{-\hbar}{m\omega}\frac{\partial^2}{\partial x^2}+\frac{m\omega}{\hbar}x^2\right]u_n(x) = \frac{2E_n}{\hbar\omega}u_n(x). \tag{5.8}$$

Introducing the variables

$$y = \left(\frac{m\omega}{\hbar}\right)^{1/2}x, \tag{5.9}$$

$$\epsilon_n = E_n/\hbar\omega, \tag{5.10}$$

the equation becomes

$$\left(\frac{\partial^2}{\partial y^2}-y^2\right)u_n(y) = -2\epsilon_n u_n(y). \tag{5.11}$$

This equation may be solved by the standard techniques, which are employed below for angular momentum and the hydrogen atom. Instead, we use the factorization method, which is particularly elegant for this problem and brings to the fore a new type of operator, which in the long run plays a very important role in the theory.

Since

$$\left(\frac{\partial}{\partial y}+y\right)\left(\frac{\partial}{\partial y}-y\right)u_n(y) \equiv \left(\frac{\partial^2}{\partial y^2}-y^2-1\right)u_n(y),$$

(5.11) may be re-written

$$\left(\frac{\partial}{\partial y}+y\right)\left(\frac{\partial}{\partial y}-y\right)u_n(y) = [-2\epsilon_n-1]u_n(y). \tag{5.11a}$$

Alternatively, it may be written

$$\left(\frac{\partial}{\partial y}-y\right)\left(\frac{\partial}{\partial y}+y\right)u_n(y) = [-2\epsilon_n+1]u_n(y). \tag{5.11b}$$

Multiply (5.11a) by $(\partial/\partial y-y)$, then

$$\left(\frac{\partial}{\partial y}-y\right)\left(\frac{\partial}{\partial y}+y\right)\left(\frac{\partial}{\partial y}-y\right)u_n(y) = [-2\epsilon_n-1]\left(\frac{\partial}{\partial y}-y\right)u_n(y). \tag{5.12}$$

Then, either

$$\left(\frac{\partial}{\partial y}-y\right)u_n(y) = 0; \tag{5.13}$$

or

$$\left(\frac{\partial}{\partial y}-y\right)u_n(y) = u_{n+1}(y), \text{ say,} \tag{5.14}$$

and (5.12) can be written

$$\left(\frac{\partial}{\partial y}-y\right)\left(\frac{\partial}{\partial y}+y\right)u_{n+1}(y) = [-2(\epsilon_n+1)+1]\,u_{n+1}(y). \quad (5.15)$$

This is (5.11b) for u_{n+1}, provided

$$\epsilon_n+1 = \epsilon_{n+1}. \quad (5.16)$$

The only solution to (5.13) is

$$u(y) = e^{+(1/2)y^2}.$$

This diverges for large y and is therefore not a possible state. Thus given any solution $u_n(y)$, eigenvalue ϵ_n, it is always possible to generate a new state $u_{n+1}(y)$ by (5.14) with eigenvalue ϵ_n+1.

Similarly, multiplying (5.11b) by $(\partial/\partial y+y)$,

$$\left(\frac{\partial}{\partial y}+y\right)\left(\frac{\partial}{\partial y}-y\right)\left(\frac{\partial}{\partial y}+y\right)u_n(y) = [-2\epsilon_n+1]\left(\frac{\partial}{\partial y}+y\right)u_n(y) \quad (5.17)$$

Now either

$$\left(\frac{\partial}{\partial y}+y\right)u_n(y) = 0; \quad (5.18)$$

or

$$\left(\frac{\partial}{\partial y}+y\right)u_n(y) = u_{n-1}(y), \text{ say}, \quad (5.19)$$

In the latter case (5.17) can be written

$$\left(\frac{\partial}{\partial y}+y\right)\left(\frac{\partial}{\partial y}-y\right)u_{n-1}(y) = [-2(\epsilon_n-1)-1]\,u_{n-1}(y), \quad (5.20)$$

which is (5.11a) for u_{n-1} provided

$$\epsilon_{n-1} = \epsilon_n-1. \quad (5.21)$$

Thus given any solution u_n, eigenvalue ϵ_n, it is possible to generate a new state of lower energy, $u_{n-1}(y)$, determined by (5.19) with eigenvalue (ϵ_n-1), unless u_n is the ground state, u_0. In this case it must satisfy (5.18);

$$\left(\frac{\partial}{\partial y}+y\right)u_0(y) = 0. \quad (5.22)$$

This determines the ground state eigenfunction to be

$$u_0(y) = e^{-(1/2)y^2}. \quad (5.23)$$

Further, by (5.22) and (5.11b), the ground state energy is

$$2\epsilon_0-1 = 0. \quad (5.24)$$

Combining these results, (5.24) and (5.16), the eigenvalues are

$$\epsilon_0 = \tfrac{1}{2}, \qquad \epsilon_1 = 1 + \tfrac{1}{2}, \dots \qquad \epsilon_n = n + \tfrac{1}{2}, \dots$$

So that

$$E_n = (\tfrac{1}{2} + n)\,\hbar\omega, \qquad n = 0, 1, 2, \dots, \tag{5.25}$$

a discrete set for all energies, in accordance with the general argument given above.

The expression for the energy levels is one of the most important in quantum mechanics. It justifies Planck's explanation of the interaction of radiation with matter, provided matter can be regarded as a collection of oscillators, each one emitting or absorbing radiation of its own frequency. The energy exchange is then restricted by the oscillator eigenvalues to take place in units of $\hbar\omega$, which is just Planck's hypothesis.

§ 5.3 The Eigenfunctions: Annihilation and Creation Operators

The successive eigenfunctions can be generated from $u_0(x)$ by repeated application of (5.14), so that, for example,

$$u_1(y) = \left(\frac{\partial}{\partial y} - y\right) u_0(y),$$

$$= \left(\frac{\partial}{\partial y} - y\right) e^{-(1/2)y^2},$$

$$= -2y\,e^{-(1/2)y^2}. \tag{5.26}$$

The ground state is an even function of y with no nodes; the first excited state is an odd function with one node. It is easy to verify by repeated applications of (5.14), that the other successive eigenfunctions have the general features derived in the previous chapter. The functions so generated are known as Hermite polynomials.

The ground state eigenfunction is in fact a Gaussian hump of the form considered in the discussion of the uncertainty principle, of width, according to (5.23) and (5.9),

$$\Delta_x = \left(\frac{\hbar}{m\omega}\right)^{1/2}. \tag{5.27}$$

According to (5.5), this is just the amplitude of the classical oscillation of the same energy as the ground state.

The operators

$$\left(\frac{\partial}{\partial y} - y\right), \quad \left(\frac{\partial}{\partial y} + y\right), \tag{5.28}$$

5

are of particular interest, since they are of a type which has not previously appeared. Up to now the operators have represented some observable (or, more pedantically, the operation of measuring some observable). A typical example is the oscillator energy operator

$$\hat{H} = \frac{\hbar\omega}{2}\left(y^2 - \frac{\partial^2}{\partial y^2}\right). \tag{5.29}$$

The above operators, however, according to (5.14) and (5.19), represent the physical operation of shifting the oscillator, up or down, from one energy level to the next. Since energy, in units of $\hbar\omega$, has to be either created or annihilated in the process, they are known respectively as creation and annihilation operators. They ultimately play a very important role in the complete quantum theory of the interaction of radiation with matter (electrons).

§ 5.4 Summary

The energy eigenvalues of a quantum harmonic oscillator of angular frequency ω are

$$E_n = (n + \tfrac{1}{2})\hbar\omega, \qquad n = 0, 1, 2, \ldots,$$

as established in (5.25).

PROBLEMS V

5.1. Use the creation operator to derive the eigenstate $u_2(x)$.

5.2. If a particle is in the state $u_0(x)$, use (3.39) to verify that there is no possibility of a measurement of the energy having the result E_1 or E_2.

5.3. Evaluate the mean value of the magnitude of the momentum

$$|p| = \left| -i\hbar\frac{\partial}{\partial x} \right|$$

for a harmonic oscillator in the ground state $u_0(x)$. (For the normalization integral, see Problems III, **3.3**.)

5.4. Define the normalized annihilation and creation operators

$$\hat{a} = \left(\frac{\hbar\omega}{2}\right)^{1/2}\left(y + \frac{\partial}{\partial y}\right),$$

$$\hat{a}^+ = \left(\frac{\hbar\omega}{2}\right)^{1/2}\left(y - \frac{\partial}{\partial y}\right).$$

Show that

$$[\hat{a}, \hat{a}^+] = \hbar\omega,$$

$$\hat{a}\hat{a}^+ = \hat{H} + \tfrac{1}{2}\hbar\omega,$$

and

$$\hat{a}^+\hat{a} = \hat{H} - \tfrac{1}{2}\hbar\omega.$$

PART II

Atomic Physics

CHAPTER 6

ANGULAR MOMENTUM

§ 6.1 Angular Momentum Operators

As a necessary preliminary to the discussion of the hydrogen atom, we now consider the quantum theory of angular momentum. It is

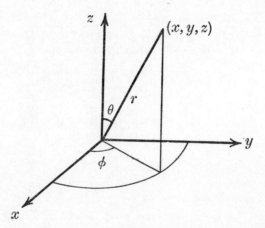

FIG. 6.1. The point (x, y, z) may also be specified by the spherical polar co-ordinates (r, θ, ϕ).

convenient to do this using spherical polar co-ordinates (r, θ, ϕ), which are related to rectangular Cartesian co-ordinates (x, y, z) by

$$x = r \sin \theta \cos \phi,$$

$$y = r \sin \theta \sin \phi,$$

$$z = r \cos \theta, \tag{6.1}$$

as is evident from Fig. 6.1. By applying the correspondence principle, P(i), the operators representing the components of angular momentum about the origin can be written in terms of those of position and linear momentum. These in turn can be expressed, by (3.14), in the Schrödinger representation. Thus

$$\hat{l}_x = \hat{y}\hat{p}_z - \hat{z}\hat{p}_y \rightarrow -i\hbar\left(y\frac{\partial}{\partial z} - z\frac{\partial}{\partial y}\right),$$

$$\hat{l}_y = \hat{z}\hat{p}_x - \hat{x}\hat{p}_z \rightarrow -i\hbar\left(z\frac{\partial}{\partial x} - x\frac{\partial}{\partial z}\right),$$

$$\hat{l}_z = \hat{x}\hat{p}_y - \hat{y}\hat{p}_x \rightarrow -i\hbar\left(x\frac{\partial}{\partial y} - y\frac{\partial}{\partial x}\right). \tag{6.2}$$

Now

$$\frac{\partial}{\partial z} = \frac{\partial r}{\partial z}\frac{\partial}{\partial r} + \frac{\partial\theta}{\partial z}\frac{\partial}{\partial\theta} + \frac{\partial\phi}{\partial z}\frac{\partial}{\partial\phi}.$$

This can be evaluated by means of (6.1), and similar expressions obtained for the other derivatives. In this way the angular momentum operators may be expressed in angular variables.

$$\hat{l}_x \rightarrow i\hbar\left(\sin\phi\frac{\partial}{\partial\theta} + \cot\theta\cos\phi\frac{\partial}{\partial\phi}\right),$$

$$\hat{l}_y \rightarrow i\hbar\left(-\cos\phi\frac{\partial}{\partial\theta} + \cot\theta\sin\phi\frac{\partial}{\partial\phi}\right),$$

$$\hat{l}_z \rightarrow -i\hbar\frac{\partial}{\partial\phi}. \tag{6.3}$$

§ 6.2 The z-component

The result for $\hat{l}_z$ could have been obtained directly from the general statement of the complementarity principle at the end of § 3.3. Since $\hat{l}_z$ is the angular momentum corresponding to the observable $\hat{\phi}$, the commutation relation of the operators is

$$[\hat{\phi},\hat{l}_z] = i\hbar.$$

If we represent the operator $\hat{\phi}$, by the algebraic variable ϕ, then for the corresponding angular momentum we have

$$\hat{l}_z(=\hat{l}_\phi) \rightarrow -i\hbar\frac{\partial}{\partial\phi}, \tag{6.4}$$

in exact analogy with (3.14).

The possible values of angular momentum are determined by the eigenvalue equation for $\hat{l}_z$,

$$-i\hbar\frac{\partial}{\partial\phi}u_m(\phi) = \hbar m u_m(\phi). \tag{6.5}$$

The eigenvalues are $\hbar m$, where the factor $\hbar$ has been extracted for convenience. Since the same physical position is denoted by $\phi + 2n\pi$ for any integer n, we must impose the boundary condition that $u_m(\phi)$ is periodic in 2π. Therefore the solutions to (6.5) are

$$u_m(\phi) = e^{im\phi} \qquad (6.6)$$

where, by the boundary condition,

$$m = 0, \pm 1, \pm 2, \ldots$$

Note, that this is just the Bohr rule of thumb, (1.20), which was imposed, on a purely *ad hoc* basis, on classical theory in order to produce the discrete energy spectrum of hydrogen in the Bohr theory. It has now appeared correctly as a particular consequence of our general theory.

§ 6.3 The Total Angular Momentum: The Eigenvalues

It is easily checked that the operator $\hat{l}_z$ does not commute with $\hat{l}_x$ and $\hat{l}_y$,

$$[\hat{l}_x, \hat{l}_z] \neq 0, \qquad [\hat{l}_y, \hat{l}_z] \neq 0. \qquad (6.7)$$

Thus, in general, an eigenstate of one component is not an eigenstate of the other components. It is consequently not possible to have an exact knowledge of more than one component, owing to the mutual disturbances of the observations. (The exceptional case is the eigenstate corresponding to the eigenvalue zero for each component, as is shown in the discussion following (6.40) below.)

However, by a further application of the correspondence principle, one can construct the operator for total angular momentum from the operators for the components,

$$\hat{l}^2 = \hat{l}_x^2 + \hat{l}_y^2 + \hat{l}_z^2,$$

$$\rightarrow -\hbar^2 \left[\frac{1}{\sin\theta} \frac{\partial}{\partial\theta} \left(\sin\theta \frac{\partial}{\partial\theta} \right) + \frac{1}{\sin^2\theta} \frac{\partial^2}{\partial\phi^2} \right]. \qquad (6.8)$$

From (6.4) and (6.8) it is simple to check that

$$[\hat{l}^2, \hat{l}_z] = 0,$$

thus it is possible for the values of the total angular momentum and z-component to be known exactly. There must exist simultaneous eigenfunctions of the two operators, $Y_{\beta m}(\theta, \phi)$, which satisfy the eigenvalue equations

$$\hat{l}_z Y_{\beta m}(\theta, \phi) = \hbar m Y_{\beta m}(\theta, \phi), \qquad (6.9)$$

$$\hat{l}^2 Y_{\beta m}(\theta, \phi) = \hbar^2 \beta Y_{\beta m}(\theta, \phi). \qquad (6.10)$$

Again we have extracted, for convenience, a factor of $\hbar^2$ from the eigenvalue, $\beta\hbar^2$, of $\hat{l}^2$. Since $\hat{l}_z$ depends only on ϕ, it must be possible to write

$$Y_{\beta m}(\theta, \phi) = P_{\beta m}(\theta) \, e^{im\phi}, \qquad (6.11)$$

which certainly satisfies (6.9). Substituting (6.11) into (6.10) gives, using (6.8),

$$\left[\frac{1}{\sin\theta}\frac{\partial}{\partial\theta}\left(\sin\theta\frac{\partial}{\partial\theta}\right) - \frac{m^2}{\sin^2\theta}\right] P_{\beta m}(\theta) = -\beta P_{\beta m}(\theta). \qquad (6.12)$$

The dependence on ϕ has been eliminated. This is an eigenvalue equation for β, which determines the possible values of $\hat{l}^2$, when $\hat{l}_z$ has the value $\hbar m$. This must be solved subject to the physically necessary boundary condition that $P_{\beta m}(\theta)$ remains finite in the physical region $0 \leqslant \theta \leqslant \pi$.

If we introduce the variable

$$w = \cos\theta, \qquad (6.13)$$

then

$$\frac{1}{\sin\theta}\frac{d}{d\theta} = -\frac{d}{dw}, \qquad (6.14)$$

and (6.12) is

$$\frac{d}{dw}(1-w^2)\frac{dP}{dw} + \left(\beta - \frac{m^2}{1-w^2}\right)P = 0, \qquad (6.15)$$

where $P(w)$ must be finite in the range

$$-1 \leqslant w \leqslant +1. \qquad (6.16)$$

For the time being we have dropped the suffix βm on P. The equation may be re-written as

$$\frac{d^2 P}{dw^2} - \frac{2w}{1-w^2}\frac{dP}{dw} + \left[\frac{\beta}{1-w^2} - \frac{m^2}{(1-w^2)^2}\right]P = 0. \qquad (6.17)$$

The singularities of the equation are at $w = \pm 1$. These are the points where the coefficients in the equation become infinite, and where P itself may become infinite, contrary to the requirements of the boundary condition.

First consider the solution near $w = 1$. We can write

$$\frac{2w}{1-w^2} = \frac{1}{1-w} - \frac{1}{1+w} \simeq \frac{1}{1-w}, \qquad (6.18)$$

and

$$\frac{1}{(1-w^2)^2} = \frac{1}{(1-w)^2(1+w)^2} \simeq \frac{1}{4(1-w)^2}, \qquad (6.19)$$

where the final approximate equalities are valid if $w \simeq 1$. For $w \simeq 1$ the β term is negligible compared with that involving m^2, and in this region the approximate equation for (6.17) is

$$\frac{d^2 P}{dw^2} - \frac{1}{1-w}\frac{dP}{dw} - \frac{m^2}{4(1-w)^2} P = 0. \tag{6.20}$$

Consider a solution of the form

$$P = (1-w)^{\alpha}[a_0 + a_1(1-w) + a_2(1-w)^2 + \ldots] \tag{6.21}$$

where, without loss of generality, one may assume

$$a_0 \neq 0. \tag{6.22}$$

Substituting (6.21) into (6.20), the coefficients of different powers of $(1-w)$ must all vanish, and in particular the vanishing of the coefficient of $(1-w)^{\alpha-2}$ gives

$$a_0\left[\alpha(\alpha-1) + \alpha - \frac{m^2}{4}\right] = 0. \tag{6.23}$$

Since a_0 is not zero, this gives

$$\alpha = \pm \frac{|m|}{2}, \tag{6.24}$$

and hence two independent solutions

$$P_0^{(1)} = (1-w)^{|m|/2}[a_0 + \ldots], \tag{6.25}$$

$$P_\infty^{(1)} = (1-w)^{-|m|/2}[a_0' + \ldots], \tag{6.26}$$

which go to zero and infinity, respectively, at $w = 1$. The boundary condition at $w = 1$ is satisfied by $P_0^{(1)}$.

By a precisely similar argument applied to the neighbourhood of $w = -1$, it can be shown that there are two independent solutions, which can be written as power series about that point;

$$P_0^{(-1)} = (1+w)^{|m|/2}[b_0 + b_1(1+w) + \ldots], \tag{6.27}$$

$$P_\infty^{(-1)} = (1+w)^{-|m|/2}[b_0' + b_1'(1+w) + \ldots]. \tag{6.28}$$

These are shown symbolically in Fig. 6.2. The boundary condition at $w = -1$ is satisfied by $P_0^{(-1)}$. In general the solution $P_0^{(-1)}$ is a linear combination of the two solutions at $w = 1$,

$$P_0^{(-1)} = aP_0^{(1)} + bP_\infty^{(1)}. \tag{6.29}$$

The required spectrum of allowed values of β are just those values for which the $P_0^{(-1)}$ solution joins smoothly onto the $P_0^{(1)}$ solution, and has no $P_0^{(1)}$ component, thus satisfying the boundary condition at both singularities. (This is mathematically analogous to the evaluation of the energy spectrum of bound states in a "square" well in Chapter 4, where the allowed values of E_n were just those for which the continuity conditions could be satisfied at the singular points of the potential, without introducing any component of the rising exponential solution.)

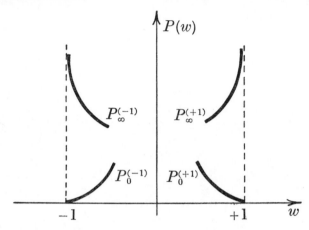

FIG. 6.2. Schematic diagram of the solutions to (6.17), showing the finite and infinite solutions at the singular points $w = \pm 1$.

To determine β we incorporate the leading terms of the $P_0^{(1)}$ and $P_0^{(-1)}$ solutions, and write

$$P_{\beta m}(w) = (1-w^2)^{|m|/2} z(w). \tag{6.30}$$

Then (6.17) is

$$(1-w^2)\frac{d^2 z}{dw^2} - 2(|m|+1)\, w\, \frac{dz}{dw} + [\beta - |m|(|m|+1)] z = 0. \tag{6.31}$$

Put

$$z(w) = \sum_0^\infty a_k w^k, \tag{6.32}$$

and equate to zero the coefficients of each power of w. The coefficient of w^k gives

$$(k+2)(k+1)a_{k+2} = [k(k-1) + 2k(|m|+1) + |m|(|m|+1) - \beta] a_k$$

$$= [(k+|m|)(k+|m|+1) - \beta] a_k. \tag{6.33}$$

This recurrence relation determines the coefficients of even powers of z from a_0, and those of the odd powers from a_1. If the series for $z(w)$ does not terminate, for large values of k (6.33) implies that

$$a_{k+2} \simeq a_k.$$

Hence, for large values of k,

$$z(w) \sim (1-w)^{-1},$$

and $P_{\beta m}$ diverges for some values of $|m|$. In order to satisfy the boundary condition the series must terminate, and $z(w)$ is therefore a polynomial, not a power series. One series of coefficients will terminate, as a result of (6.33), at the kth term if

$$\beta = l(l+1), \tag{6.34}$$

where

$$l = k + |m|. \tag{6.35}$$

If

$$k = l - |m|$$

is odd, and if $a_0 = 0$, the corresponding solution $P_l^m(w)$ is a polynomial of odd powers. If k is even, and if $a_1 = 0$, $P_l^m(w)$ is a polynomial of even powers. These polynomials, $P_l^m(w)$, are, in fact, well known to mathematicians as Associated Legendre Polynomials. With these eigenfunctions the boundary conditions are satisfied and (6.34) determines the allowed values of β.

The eigenvalues of $\hat{l}^2$, determined by (6.34), are

$$\hbar^2 \beta = \hbar^2 l(l+1), \qquad l = 0, 1, 2\ldots, \tag{6.36}$$

since, by (6.35), l is greater than or equal to $|m|$. (The states with l values $0, 1, 2, 3, 4, \ldots$ are known for historical reasons as $S, P, D, F, G, \ldots$ states, respectively.)

For any given value of l, the possible eigenvalues of $\hat{l}_z$, are

$$l_z = \hbar m, \qquad m = 0, \pm 1, \pm 2, \ldots, \pm l. \tag{6.37}$$

giving $(2l+1)$ allowed values.

§ 6.4 The Eigenfunctions and the Vector Diagram

This result may be compared with the classical one. If angular momentum is measured in units of $\hbar$, the quantum eigenvalue $l(l+1)$ corresponds approximately to a classical angular momentum of magnitude l. The total angular momentum vector can point in any direction, which can be specified by the angles θ, and ϕ. Its z-component is independent of ϕ, but clearly depends on θ. It has its maximum value, l, when $\theta = 0$; and its minimum value, $-l$, when $\theta = \pi$. Classically l_z can take on any value intermediate between these two extremes, depending on the precise orientation with respect to θ. The quantum mechanical result is qualitatively similar, except that the allowed values are now only the integers lying between the extreme values, $\pm l$.

In classical terms this may be represented by the vector diagram of Fig. 6.3, in which the allowed values of l_z are shown as allowed

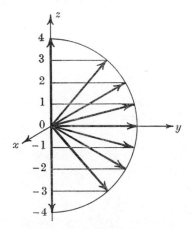

Fig. 6.3. The vector diagram of angular momentum. Classically the angular momentum vector may have any orientation. Quantum mechanically it is restricted to values for which the z-component is an integer multiple of $\hbar$.

orientations (particular values of θ) for the angular momentum vector. This model should not be taken too seriously since for states of definite angular momentum, there must always be some uncertainty in the orientation since, loosely speaking, these are complementary variables.

If the physical system consists of a particle moving in some central potential about the origin of co-ordinates, the probability distribution of its orientation for given angular momentum states is correctly given by the square modulus of the angular momentum eigenstates, which we can now write as $Y_l^m(\theta, \phi)$. These are known as Spherical Harmonics. If they are normalized so that the total probability of finding the particle with some orientation is unity,

$$Y_l^m(\theta, \phi) = i^{m+|m|} \left[\frac{2l+1}{4\pi} \frac{(l-|m|)!}{(l+|m|)!} \right]^{1/2} P_l^m(\cos \theta) \, e^{im\phi}. \quad (6.38)$$

For $m = 0$ we have the relations

$$\int_{-1}^{+1} P_l(w) \, P_{l'}(w) \, dw = \frac{2}{2l+1} \, \delta_{ll'}, \qquad (6.39)$$

$$P_l(1) = 1, \qquad P_l(-1) = (-1)^l. \qquad (6.40)$$

For the simple cases, (6.38) implies the explicit expressions shown in Table 6.1. Note that Y_0^0 is a constant and is, in fact, also an eigenstate of $\hat{m}_x$ and $\hat{m}_y$, corresponding to the eigenvalue zero for each. This is the exceptional case anticipated in § 6.3.

TABLE 6.1. *The angular momentum eigenstates for S and P waves*

l	$m = 1$	$m = 0$	$m = -1$
0		$Y_0^0 = \left(\dfrac{1}{4\pi}\right)^{1/}$	
1	$Y_1^{+1} = -\left(\dfrac{3}{4\pi}\right)^{1/2} \sin\theta \dfrac{e^{i\phi}}{\sqrt{2}};$	$Y_1^0 = \left(\dfrac{3}{4\pi}\right)^{1/2} \cos\theta;$	$Y_1^{-1} = \left(\dfrac{3}{4\pi}\right)^{1/2} \sin\theta \dfrac{e^{-i\phi}}{\sqrt{2}}$

For a state specified by l, m, the probability of the particle having an orientation θ, ϕ is, by (3.38),

$$\mathscr{P}_{l,m}(\theta, \phi) = |Y_l^m(\theta, \phi)|^2. \qquad (6.41)$$

For the states $l = 1$, $m = \pm 1$, it is clear from Table 6.1 that the most probable values for θ is $\pi/2$, so that the "orbit" tends to lie in the x-y plane, and the total angular momentum vector consequently is most probably up or down the z-axis. On the other hand for $l = 1$, $m = 0$, the favoured values of θ are 0 and π, so that the "orbit" tends to lie perpendicular to the x-y plane and the angular momentum vector is most probably in the x-y plane. There is no physical dependence on ϕ. This shows the sense in which the semi-classical vector diagram (Fig. 6.3) should be understood in terms of the correct quantum mechanical formalism.

In the classical limit the values of l and m become very large, the distinction between l and $[l(l+1)]^{1/2}$ is completely negligible, as is also the difference between the continuous and allowed discrete spectrum of values for m. One thus goes over into the ordinary classical picture.

§ 6.5 Parity

In discussing the bound states of a system in a one-dimensional symmetric potential (Chapter 4), it proved convenient to introduce the notion of parity. It is now convenient to extend this to three dimensions, in terms of polar co-ordinates. Suppose the position of a particle is specified with respect to conventionally defined polar

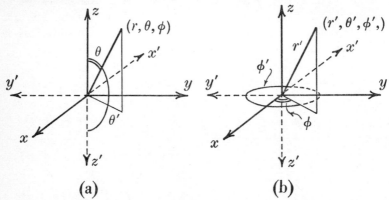

(a)   **(b)**

FIG. 6.4. Spherical polar co-ordinates of the point (r, θ, ϕ) in a reflected co-ordinate frame. r', θ' and ϕ' are defined in the conventional way, but with respect to x', y' and z'.

variables by co-ordinates (r, θ, ϕ). If the axes are reflected to (x', y', z'), and r', θ' and ϕ' are defined in the usual way, but with respect to x', y', and z', then from Fig. 6.4 it can be seen that

$$r' = r,$$
$$\theta' = \pi - \theta,$$
$$\phi' = \pi + \phi. \tag{6.42}$$

Thus if a particle is in an angular momentum state $Y_l^m(\theta, \phi)$, according to the reflected axes it is in a state

$$
\begin{aligned}
Y_l^m(\theta', \phi') &= \text{const. } P_l^m[\cos \theta'] e^{im\phi'} \\
&= \text{const. } P_l^m[\cos(\pi - \theta)] e^{im(\pi + \phi)} \\
&= \text{const. } P_l^m(-\cos \theta) e^{im\phi}(-1)^{|m|} \\
&= \text{const. } (-1)^{l-|m|} P_l^m(\cos \theta) e^{im\phi}(-1)^{|m|}, \quad (6.43)
\end{aligned}
$$

where we have used (6.42) and, to obtain the final equality, the remarks following (6.35). Combining, we have the result that

$$Y_l^m(\theta', \phi') = (-1)^l Y_l^m(\theta, \phi). \tag{6.44}$$

Since, by (6.42), the dependence of any state function on r is unchanged by reflection, the parity of any state of definite angular momentum l, m is determined by the l-value only, and is equal to $(-1)^l$.

§ 6.6 Summary

The main conclusions of this chapter are that the eigenvalues of the total angular momentum $\hat{l}^2$ are

$$\hbar^2 l(l+1), \qquad l = 0, 1, 2, \ldots$$

and those of its z-component $\hat{l}_z$ are

$$m\hbar, \qquad m = 0, \pm 1, \ldots \pm l.$$

PROBLEMS VI

6.1. Use the representations (6.2) to verify the operator equation

$$[\hat{l}_x, \hat{l}_y] = i\hbar \hat{l}_z.$$

(This has important consequences, particularly in connection with spin, § 8.3.)

6.2. A rigid system rotates freely about the z-axis with moment of inertia I. By expressing the energy of the system in terms of the angular momentum, $\hat{l}_z$, show that the possible energy levels of the system are

$$E_m = \frac{\hbar^2 m^2}{2I}, \qquad m = 0, 1, 2, \ldots$$

with eigen functions

$$u_m(\phi) = e^{\pm im\phi},$$

where ϕ is the angle specifying the orientation of the system in the x–y plane.

6.3. The rotational energy of a diatomic molecule may be obtained by regarding the molecule as a rigid system consisting of two point particles a fixed distance apart, which is free to rotate about its centre of gravity. By expressing the total energy in terms of the moment of inertia, I, and the total angular momentum, show that the rotational energy levels are

$$E_l = \frac{\hbar^2}{2I} l(l+1), \qquad l = 0, 1, 2, \ldots$$

and that the level specified by l has $(2l+1)$ different eigenstates, corresponding to the possible values of the z-component of angular momentum

$$l_z = \hbar m, \qquad m = 0, \pm 1, \pm 2, \ldots \pm l.$$

6.4. If the system of Problem **6.3** is in a state

$$\psi(\theta, \phi) = \frac{1}{\sqrt{3}} Y_1^0(\theta, \phi) + \sqrt{\left(\frac{2}{3}\right)} Y_1^1(\theta, \phi),$$

show that the energy of the state is E_1, but that (after summing over all possible values of ϕ) the probability distribution of the orientation, θ, is the same as for a particle in the ground state

$$u_{0,0}(\theta, \phi) = Y_0^0(\theta, \phi).$$

(Note that the probability of finding the system oriented in a solid angle $d\Omega(\theta, \phi)$ is

$$|\psi(\theta, \phi)|^2 d\Omega(\theta, \phi) = |\psi(\theta, \phi)|^2 \sin\theta \, d\theta \, d\phi.)$$

CHAPTER 7

CENTRAL POTENTIAL: THE HYDROGEN ATOM

§ 7.1 Motion in a Central Potential

The energy operator (Hamiltonian) for a particle of mass m_e in an arbitrary central potential $V(r)$ is

$$\hat{H} \rightarrow \left[\frac{-\hbar^2}{2m_e}\nabla^2 + V(r)\right], \tag{7.1}$$

and, according to (3.18), the energy eigenvalue equation is

$$\left[\frac{-\hbar^2}{2m_e}\nabla^2 + V(r)\right]u_{E_n}(r,\theta,\phi) = E_n u_{E_n}(r,\theta,\phi). \tag{7.2}$$

In spherical polar co-ordinates this is

$$\left[\frac{-\hbar^2}{2m_e}\left(\frac{1}{r^2}\frac{\partial}{\partial r}\left(r^2\frac{\partial}{\partial r}\right) + \frac{1}{r^2\sin\theta}\frac{\partial}{\partial\theta}\left(\sin\theta\frac{\partial}{\partial\theta}\right) + \frac{1}{r^2\sin^2\theta}\frac{\partial^2}{\partial\phi^2}\right) + \right.$$

$$\left. + V(r)\right]u_{E_n}(r,\theta,\phi) = E_n u_{E_n}(r,\theta,\phi). \tag{7.3}$$

The dependence on θ, ϕ is precisely that of the total angular momentum operator $\hat{l}^2$ (see (6.8)) so this can be written

$$\left[\frac{-\hbar^2}{2m_e}\frac{1}{r^2}\frac{\partial}{\partial r}\left(r^2\frac{\partial}{\partial r}\right) + \frac{\hat{l}^2(\theta,\phi)}{2m_e r^2} + V(r)\right]u_{E_n}(r,\theta,\phi) = E_n u_{E_n}(r,\theta,\phi). \tag{7.4}$$

In this form it is easy to see that

$$[\hat{H}, \hat{l}^2] = 0, \tag{7.5}$$

and also, from (6.4), that

$$[\hat{H}, \hat{l}_z] = 0. \tag{7.6}$$

Thus, it is possible to know simultaneously the energy, total angular momentum, and z-component of angular momentum for a particle moving in a central potential. If the two latter variables are specified by l and m, as in Chapter 6, the corresponding eigenfunction can be specified as u_{nlm}. Since the θ, ϕ dependence of $\hat{H}$ is contained entirely in the term $\hat{l}^2$, one can write

$$u_{nlm}(r,\theta,\phi) = u_{nl}(r)\, Y_l^m(\theta,\phi). \tag{7.7}$$

6

Substituting (7.7) into (7.4), the effect of $\hat{l}^2$ on $Y_l^m(\theta,\phi)$ is to multiply it by $\hbar^2 l(l+1)$. There is no further dependence on θ and ϕ, so the factor $Y_l^m(\theta,\phi)$ can be dropped on both sides of the equation, leaving

$$\left[\frac{-\hbar^2}{2m_e}\frac{1}{r^2}\frac{\partial}{\partial r}\left(r^2\frac{\partial}{\partial r}\right)+\frac{\hbar^2 l(l+1)}{2m_e r^2}+V(r)\right]u_{nl}(r) = E_n u_{nl}(r). \quad (7.8)$$

If one puts, in (7.8),

$$u_{nl}(r) = \frac{\chi_{nl}(r)}{r}, \quad (7.9)$$

a little straightforward algebra is sufficient to show that

$$\left[\frac{-\hbar^2}{2m_e}\frac{\partial^2}{\partial r^2}+\frac{\hbar^2}{2m_e}\frac{l(l+1)}{r^2}+V(r)\right]\chi_{nl}(r) = E_n \chi_{nl}(r). \quad (7.10)$$

In this form the equation has a simple physical interpretation.

$$\frac{-\hbar^2}{2m_e}\frac{\partial^2}{\partial r^2} \to \frac{p_r^2}{2m_e}, \quad (7.11)$$

where p_r is the radial momentum. If p_t is the transverse momentum, then the square of the total angular momentum is

$$\hbar^2 l(l+1) = (p_t r)^2. \quad (7.12)$$

Thus

$$\frac{\hbar^2 l(l+1)}{2m_e r^2} = \frac{p_t^2}{2m_e}, \quad (7.13)$$

and the whole Hamiltonian is

$$H = \frac{p_r^2}{2m_e}+\frac{p_t^2}{2m_e}+V(r). \quad (7.14)$$

§ 7.2 The Hydrogen Atom

To discuss the energy levels of the hydrogen atom we use the eigen-value equation in the form (7.8) with the Coulomb potential, which we take to be

$$V(r) = \frac{-Ze_M^2}{r}. \quad (7.15)$$

Here Ze is the charge on the nucleus. Thus

$$\left[\frac{-\hbar^2}{2m_e}\frac{1}{r^2}\frac{\partial}{\partial r}\left(r^2\frac{\partial}{\partial r}\right)-\frac{Ze_M^2}{r}+\frac{l(l+1)\hbar^2}{2m_e r^2}\right]u_{nl}(r) = E_n u_{nl}(r). \quad (7.16)$$

The algebra required to solve this equation is tedious, but the result is so important that we sketch the main steps. The method is identical with that employed in Chapter 6 to find the eigenvalues of l^2. Since we are interested in the bound states, put

$$E_n = -|E_n|,$$

and introduce the variables

$$\alpha_n^2 = \frac{8m_e|E_n|}{\hbar^2}, \qquad \rho = \alpha_n r, \qquad \lambda_n = \frac{Ze_M^2}{\hbar}\left(\frac{m_e}{2|E_n|}\right)^{1/2}. \quad (7.17)$$

Then the equation is

$$\left[\frac{1}{\rho^2}\frac{d}{d\rho}\left(\rho^2\frac{d}{d\rho}\right) + \frac{\lambda_n}{\rho} - \frac{1}{4} - \frac{l(l+1)}{\rho^2}\right]u_{nl}(\rho) = 0. \quad (7.18)$$

The singularities of this equation are at $\rho = 0$ and $\rho = \infty$. It has to be solved for the eigenvalues, λ_n, for which it is possible to find solutions which are finite everywhere, particularly at the singularities.

Consider large ρ. The equation which approximates to (7.18) is

$$\left(\frac{d^2}{d\rho^2} - \frac{1}{4}\right)u(\rho) = 0. \quad (7.19)$$

There are thus two independent solutions for large ρ.

$$u(\rho) \simeq \rho^s e^{\pm\rho/2}, \quad (7.20)$$

and the boundary condition requires the falling exponential. To discuss the behaviour at the origin consider a solution of the form

$$u_{nl}(\rho) = \rho^s e^{-\rho/2} L_{nl}(\rho). \quad (7.21)$$

Then substituting (7.21) into (7.18) gives (dropping the suffixes n, l on L for the moment),

$$\left[\rho^2\frac{d^2}{d\rho^2} + \rho\{2(s+1) - \rho\}\frac{d}{d\rho} + \{\rho(\lambda_n - s - 1) + s(s+1) - l(l+1)\}\right]L(\rho)$$
$$= 0. \quad (7.22)$$

Putting $\rho = 0$,

$$s(s+1) = l(l+1). \quad (7.23)$$

Therefore, either

$$s = l, \quad (7.24)$$

or

$$s = -(l+1), \quad (7.25)$$

which again gives two independent solutions at the origin, of which only that given by $s = l$ is allowed by the boundary condition. The situation is illustrated in Fig. 7.1 and is exactly analogous to that for the eigenvalues of $\hat{l}^2$ discussed in Chapter 6. In general, the solution corresponding to $s = l$ joins smoothly to a linear combination of both $e^{\rho/2}$ and $e^{-\rho/2}$. The eigenvalues of λ_n, and hence E_n, are those particular values for which no component of the $e^{\rho/2}$ solution is present.

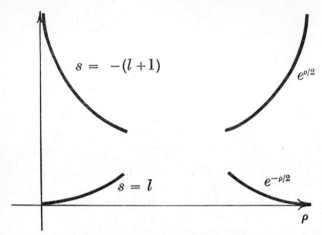

FIG. 7.1. Schematic diagram of the solutions to (7.18), showing the finite and infinite solutions at the singularities $\rho = 0$ and $\rho = \infty$.

To find these, consider (7.22) with $s = l$,

$$\left[\rho\frac{d^2}{d\rho^2} + \{2(l+1) - \rho\}\frac{d}{d\rho} + (\lambda_n - l - 1)\right] L(\rho) = 0. \qquad (7.26)$$

Substitute the power series for $L(\rho)$,

$$L(\rho) = \sum_\nu a_\nu \rho^\nu, \qquad (7.27)$$

and equate to zero the coefficient of ρ^ν. Then

$$(\nu+1)[\nu+2(l+1)]a_{\nu+1} = [\nu+(l+1-\lambda_n)]a_\nu. \qquad (7.28)$$

If the series does not terminate, for large ν

$$a_{\nu+1} \simeq \frac{1}{\nu}a_\nu, \qquad (7.29)$$

so that

$$L(\rho) \simeq e^\rho, \qquad (7.30)$$

and hence

$$u(\rho) \simeq e^\rho e^{-\rho/2}, \qquad (7.31)$$

which is not allowed. Thus to satisfy the boundary condition the series must terminate. This is ensured by (7.28) if λ_n is an integer,

$$\lambda_n = n(> l). \tag{7.32}$$

In this case, by (7.17),

$$E_n = -\frac{m_e Z^2 e_M^4}{2\hbar^2}\frac{1}{n^2} = -\frac{1}{2}\frac{Z^2 e_M^2}{a_0}\left(\frac{1}{n^2}\right), \tag{7.33}$$

where a_0 is the Bohr radius, defined in (1.24). For $Z = 1$ this is the correct formula for the Bohr levels of hydrogen.

§ 7.3 The Quantum Numbers

The energy levels of hydrogen are completely specified by a single quantum number, the *principal quantum number*

$$n = 1, 2, 3, \ldots$$

For a given energy, specified by n, just as in classical mechanics, there is a range of possible values for the total angular momentum. Classically this range varies continuously from zero, corresponding to the extremely eccentric elliptic orbit, which reduces to a linear oscillation through the origin, to a circular orbit whose radius is fixed by the energy. Quantum mechanically the range is essentially the same, but only certain discrete values are allowed, determined by l, *the orbital quantum number*, where, by (7.32),

$$l = 0, 1, 2, \ldots, n-1,$$

and the eigenvalue of the square of the angular momentum is

$$\hbar^2 l(l+1).$$

Given l, there is then a range of discrete values of the z-component of angular momentum, which partially determines the orientation of the angular momentum vector in a manner already discussed in Chapter 6. This is specified by the *magnetic quantum number*,

$$m = 0, \pm 1, \pm 2, \ldots, \pm l; \qquad (2l+1) \text{ values,}$$

$$l_z = \hbar m.$$

(The name arises because levels with different m values may be split if the atom lies in an external magnetic field. See § 8.1.)

The three numbers n, l, m determine a unique eigenfunction. Since there are a number of independent eigenfunctions for each energy level, the levels are said to be degenerate. The degree of degeneracy is the number of eigenstates corresponding to a particular level, which for the nth level is

$$D_n = \sum_{l=0}^{n-1} (2l+1) = n^2.$$

The interrelation of the quantum numbers n, l, m, combined with the ideas of intrinsic spin and the exclusion principle, forms the physical basis of the Periodic Table of the chemical elements, which is discussed in § 8.5.

§ 7.4 The Eigenfunctions

The normalized eigenfunction specified by n, l, m is

$$u_{nlm}(r, \theta, \phi) = N_{nl} u_{nl}(r) Y_l^m(\theta, \phi), \qquad (7.34)$$

where N_{nl} is a normalizing factor. The radial dependence in terms of ρ, (see 7.17), is

$$u_{nl}(r) = e^{-\rho/2} \rho^l L_{nl}(\rho). \qquad (7.35)$$

The polynomials L are the Associated Laguerre functions and are usually written

$$L_{n+l}^{2l+1}(\rho).$$

The dominating radial dependence comes from the exponential factor. Substituting (7.33) into (7.17),

$$\alpha_n = \frac{2Z}{a_0 n}, \qquad \rho = \alpha_n r. \qquad (7.36)$$

Therefore, dropping the suffix on a_0 for the rest of this section,

$$u_n(r) \sim e^{-\rho/2} = \exp\left[\frac{-Zr}{an}\right]. \qquad (7.37)$$

For the hydrogen ground state, $(Z = 1, n = 1)$, this gives a falling exponential, $e^{-r/a}$, which approximately confines the particle to the classically allowed region of $r < a$ (see Fig. 3.1), though there is an appreciable penetration of the potential barrier by the tail of the exponential distribution. As the excitation increases, n increases, and the particle is likely to be found over a wider range of r. This is in accordance with the widening of the potential, and consequently of the classically allowed region, as the energy increases.

If Z increases, the particle is confined to a smaller region as one would expect. Also, since a is inversely proportional to m_e (the mass of the bound particle), a heavier particle is bound in a smaller region. Thus if the electron is replaced by a μ-meson (see Table 11.1), the scale of the whole system is reduced by the ratio of the electron and μ-meson masses, which is a factor of over 200. Two simple normalized eigenfunctions are

$$u_{100} = \frac{1}{\pi^{1/2}}\left(\frac{Z}{a}\right)^{3/2} \exp\left[\frac{-Zr}{a}\right], \tag{7.38}$$

$$u_{200} = \frac{1}{\pi^{1/2}}\left(\frac{Z}{2a}\right)^{3/2}\left(1 - \frac{Zr}{2a}\right)\exp\left[\frac{-Zr}{2a}\right]. \tag{7.39}$$

Thus for these states the probability of finding the particle at the origin varies as a^{-3} or m_e^3, where m_e is the mass of the bound particle.

§ 7.5 Centre of Mass Motion

We have so far treated the nucleus as a fixed source about which the electron moves. In fact there is a mutual interaction between the nucleus and the electron, and the combined system is free to move. Classically the motion of such a system can be reduced to the free motion of the total mass of the system, concentrated at the centre of mass, and the relative motion, which is equivalent to that of a particle of "reduced" mass moving in a fixed potential. We now show that this is also true of quantum mechanical systems.

Consider two particles with co-ordinates and masses (x_1, y_1, z_1), m_1 and (x_2, y_2, z_2), m_2, respectively, moving in a potential which depends only on the distance between them. The Hamiltonian is

$$\hat{H} = -\frac{\hbar^2}{2m_1}\nabla_{(1)}^2 - \frac{\hbar^2}{2m_2}\nabla_{(2)}^2 + V(\mathbf{x}_1 - \mathbf{x}_2), \tag{7.40}$$

and the equation for the energy levels is

$$\hat{H}(\mathbf{x}_1, \mathbf{x}_2)\, U(\mathbf{x}_1, \mathbf{x}_2) = E''\, U(\mathbf{x}_1, \mathbf{x}_2). \tag{7.41}$$

Introduce the variables

$$M = m_1 + m_2, \tag{7.42}$$

and

$$m_1 x_1 + m_2 x_2 = MX, \text{ etc.}, \tag{7.43}$$

$$x_1 - x_2 = x, \text{ etc.}, \tag{7.44}$$

so that (X, Y, Z) are the centre of mass co-ordinates, and (x, y, z) the relative co-ordinates. Then

$$\frac{\partial}{\partial x_1} = \frac{\partial x}{\partial x_1}\frac{\partial}{\partial x} + \frac{\partial X}{\partial x_1}\frac{\partial}{\partial X} \tag{7.45}$$

$$= \frac{\partial}{\partial x} + \frac{m_1}{M}\frac{\partial}{\partial X}, \tag{7.46}$$

and similarly

$$\frac{\partial}{\partial x_2} = -\frac{\partial}{\partial x} + \frac{m_2}{M}\frac{\partial}{\partial X}.$$ (7.47)

It is then simple to show that

$$\frac{1}{m_1}\nabla^2_{(1)} + \frac{1}{m_2}\nabla^2_{(2)} = \frac{1}{\mu}\nabla^2_x + \frac{1}{M}\nabla^2_X,$$ (7.48)

where μ is the reduced mass,

$$\frac{1}{\mu} = \frac{1}{m_1} + \frac{1}{m_2}.$$ (7.49)

The Schrödinger equation, (7.41), in the new variables is thus

$$\left[-\frac{\hbar^2}{2M}\nabla^2_X - \frac{\hbar^2}{2\mu}\nabla^2_x + V(\mathbf{x})\right]U(\mathbf{x}, \mathbf{X}) = E'' U(\mathbf{x}, \mathbf{X}).$$ (7.50)

Since there are no terms in the square bracket which depend on both $\mathbf{x}$ and $\mathbf{X}$, we may look for a solution of the form

$$U(\mathbf{x}, \mathbf{X}) = u(\mathbf{x})\,w(\mathbf{X}).$$ (7.51)

Substituting into (7.50), and dividing by $u(\mathbf{x})w(\mathbf{X})$, gives

$$\frac{1}{w(\mathbf{X})}\left[\frac{-\hbar^2}{2M}\nabla^2_X\right]w(\mathbf{X}) + \frac{1}{u(\mathbf{x})}\left[\frac{-\hbar^2}{2\mu}\nabla^2_x + V(\mathbf{x})\right]u(\mathbf{x}) = E''.$$ (7.52)

The two terms on the left-hand side may be varied independently by varying either $\mathbf{X}$, or $\mathbf{x}$, respectively. They must thus each be equal to a constant. The equation splits into

$$\frac{1}{w(\mathbf{X})}\left[\frac{-\hbar^2}{2M}\nabla^2_X\right]w(\mathbf{X}) = E'' - E = E',$$ (7.53)

and

$$\frac{1}{u(\mathbf{x})}\left[\frac{-\hbar^2}{2\mu}\nabla^2_x + V(\mathbf{x})\right]u(\mathbf{x}) = E.$$ (7.54)

Both these equations have been met before. The first is simply the energy eigenvalue equation for the motion of a free particle of mass M. The energy spectrum is continuous, so that, as in the classical case, the centre of mass of the system moves freely, and with any energy. The second equation is the eigenvalue equation for the energy of the relative motion. As in the classical case, it is formally equivalent to the equation for a particle of the reduced mass μ, moving in a fixed potential.

If V is taken to be the Coulomb potential, it is the eigenvalues determined by (7.54) which are the actual energy levels of hydrogen, allowing for the motion of the nucleus. The solution is obtained trivially from (7.33) by the replacement

$$m_e \to \mu = \frac{m_e m_p}{m_e + m_p},$$

where m_p is the proton mass. This is a small correction which is, however, well within experimental errors (see § 8.6).

This analysis into centre of mass, and relative motion is absolutely essential to the consideration of particle collisions (scattering), discussed below in Chapter 10. It is evident from the definition, (7.49), that the reduced mass, μ, is equal to one of the particle masses, say m_1, when the other particle mass, m_2, becomes infinite. In general, the fixed source approximation will be a good one whenever one particle (the target particle in a two-particle collision) is very massive compared to the other particle (the projectile).

§ 7.6 General Remarks

The derivation of the formula (7.33) for the levels of hydrogen is proof that the formalism which we have set up for dealing with the mechanics of quantum systems is essentially correct. The impact of the result on the reader is greatly reduced, since the final answer is already familiar from the semi-classical Bohr picture. It is worth repeating that the Bohr picture is a completely inconsistent one, based on the arbitrary superposition of rules of thumb on classical mechanics, which are quite contrary to the spirit of classical physics. We have now put the formula on a firm physical foundation, by basing it on a completely consistent theory of the physics of small systems, appreciably disturbed by the most delicate observations.

Having re-derived this formula on this basis, one can now apply quantum mechanics confidently and unambiguously in any field in which Planck's constant, $\hbar$, cannot be neglected. This includes the whole of atomic spectroscopy; in principle, the whole of chemistry; the detailed structure of solids including their elastic and their electric and magnetic properties; and even, to the internal workings of the atomic nuclei, as we show below.

PROBLEMS VII

7.1. The relative probability distribution of the magnitude of the momentum in the ground state of hydrogen is

$$\mathscr{P}_{u_{100}}(p) = |\phi(p)|^2,$$

where, by (3.39) and (7.38),

$$\phi(p) = \iiint \exp[-i\mathbf{p}.\mathbf{r}/\hbar]\exp[-r/a]r^2\sin\theta\,dr\,d\theta\,d\phi.$$

Show that

$$\phi(p) \sim \frac{1}{[p^2 + (\hbar/a)^2]^2}.$$

(*Hint.* To evaluate the integral take the z-axis of $\mathbf{r}$ in the direction of $\mathbf{p}$. The integral over the angles is then identical with the one carried out in the derivation of (10.31). The remaining integral over r can be carried out "by parts".)

CHAPTER 8

SPIN AND STATISTICS

§ 8.1 The Zeeman Effect

Consider a hydrogen atom in a constant weak magnetic field B. To discuss the quantum effects of such a field, we must add to the Hamiltonian operator the term representing the interaction energy between the atom and the magnetic field. According to the correspondence principle, P(i), (§ 3.3), this is formally the same as the classical expression. To derive it, it is sufficient to consider the proton as fixed and the electron moving around it in a circular classical orbit. The electron orbit can then be regarded as a current loop, which is equivalent to a magnetic dipole of moment μ, (μ being a vector perpendicular to the plane of the loop). The interaction energy is

$$V_B = -\mu . B. \tag{8.1}$$

If the radius of the orbit is r, and p the momentum, the velocity of the electron in its orbit is p/m_e. The current is the charge per unit time past any fixed point on the orbit, which is†

$$j = -\frac{e(p/m_e)}{2\pi r}. \tag{8.2}$$

The magnitude of the magnetic moment of the loop is the current times the enclosed area. Thus

$$|\mu| = \pi r^2 j \tag{8.3}$$

$$= \frac{-erp}{2m_e}. \tag{8.4}$$

Therefore

$$\mu = \frac{-e}{2m_e} r \wedge p = \frac{-e}{2m_e} l, \tag{8.5}$$

where l is the angular momentum. This is a general result, not restricted to circular orbits. The interaction energy is then

$$V_B = \frac{e}{2m_e} B . l. \tag{8.6}$$

† To transform to Gaussian units replace B by H, j by j/c and e by e/c throughout § 8.1.

If we now interpret l as the angular momentum operator, this is the correct quantum mechanical expression.

If we take the z-axis in the direction of B, the total Hamiltonian for the system is

$$\hat{H}_B = \hat{H} + \frac{e}{2m_e} B \hat{l}_z, \tag{8.7}$$

where $\hat{H}$ is the Hamiltonian for an undisturbed atom given in (7.16). The energy levels of the atom in the field B are determined by the eigenvalue equation for this new Hamiltonian,

$$\hat{H}_B u_n(r, \theta, \phi) = E_n^B u_n(r, \theta, \phi). \tag{8.8}$$

The extra term in $\hat{H}_B$ only contains the operator $\hat{l}_z$. The eigenstates for $\hat{H}$ are already eigenstates of this operator. The eigenstates u_n of $\hat{H}_B$ are therefore the same as for $\hat{H}$, namely u_{nlm} of § 7.4. Thus

$$\hat{H}_B u_{nlm}(r, \theta, \phi) = \left[\hat{H} + \frac{e}{2m_e} B \hat{l}_z \right] u_{nlm}(r, \theta, \phi)$$

$$= \left(E_n + \frac{e}{2m_e} B \hbar m \right) u_{nlm}(r, \theta, \phi), \tag{8.9}$$

where E_n are the unperturbed hydrogen levels, (7.33). For given values of n and l, the $2l+1$ different states, which previously corresponded to the same level, E_n, are now split into $2l+1$ separate levels with spacing, according to (8.9), of

$$\Delta E = \frac{e \hbar B}{2m_e}. \tag{8.10}$$

This effect is observed, and shows why m is known as the magnetic quantum number (§ 7.3).

According to (8.9) there should be no splitting of the ground state, $(n = 1, l = 0, m = 0)$. However, experimentally, the lowest level is observed to split into two. If this splitting has the same physical origin as that derived above, it must be associated with an angular momentum j, which satisfies

$$2j+1 = 2.$$

Therefore

$$j = \tfrac{1}{2}. \tag{8.11}$$

Since we have shown quite generally, in Chapter 6, that orbital angular momentum can only take on integer values, this result indicates the necessity for some generalization of the formalism. The answer is to introduce a new type of operator—matrix operators.

§ 8.2 Matrix Operators

A matrix is an $(n \times n)$ array of numbers, where n may have any value. We deal explicitly with the simplest non-trivial case, $n = 2$. Thus†

$$\hat{A} = \begin{pmatrix} \langle 1|\hat{A}|1\rangle & \langle 1|\hat{A}|2\rangle \\ \langle 2|\hat{A}|1\rangle & \langle 2|\hat{A}|2\rangle \end{pmatrix}. \tag{8.12}$$

The numbers $\langle i|\hat{A}|j\rangle$, $i = 1,\dots,n$, $j = 1,\dots,n$, are called the elements of the matrix. The label $\langle i|$ denotes the row, and $|j\rangle$ the column in which the element appears. The matrix operates on a "vector", which is a column of n numbers (or components), which we write (for $n = 2$)†,

$$|\psi\rangle = \begin{pmatrix} \langle 1|\psi\rangle \\ \langle 2|\psi\rangle \end{pmatrix}. \tag{8.13}$$

When a matrix $\hat{A}$ operates on a vector $|\psi\rangle$, it produces a new vector $|\phi\rangle$;

$$\hat{A}|\psi\rangle = |\phi\rangle. \tag{8.14}$$

The components of $|\phi\rangle$ are determined by the rule

$$\sum_{j=1}^{n} \langle i|\hat{A}|j\rangle\langle j|\psi\rangle = \langle i|\phi\rangle, \qquad i = 1,\dots,n. \tag{8.15}$$

An eigenvalue for the matrix operator, $\hat{A}$, can be defined, in exact analogy with (2.7), by

$$\hat{A}|u_n\rangle = a_n|u_n\rangle, \tag{8.16}$$

where a_n and $|u_n\rangle$ are the eigenvalues and eigenvectors of $\hat{A}$, respectively.

If the matrix $\hat{A}$ is of the diagonal form

$$\hat{A} = \begin{pmatrix} a_1 & 0 \\ 0 & a_2 \end{pmatrix}, \tag{8.17}$$

it is very simple to check by substitution into (8.16) that the eigenvalues are a_1 and a_2, with eigenvectors

$$|u_1\rangle = \begin{pmatrix} 1 \\ 0 \end{pmatrix}, \qquad |u_2\rangle = \begin{pmatrix} 0 \\ 1 \end{pmatrix}. \tag{8.18}$$

† The alternative notation

$$\langle i|\hat{A}|j\rangle = A_{ij}$$
$$\langle i|\psi\rangle = \psi_i \text{ is also frequently used.}$$

The product of a matrix $\hat{A}$ with a number c, $c\hat{A}$, implies a new matrix, each element of which is c times the corresponding element of $\hat{A}$.

$$\langle i|c\hat{A}|j\rangle = c\langle i|\hat{A}|j\rangle. \tag{8.19}$$

The product of two $(n \times n)$ matrices $\hat{A}$ and $\hat{B}$ is also an $(n \times n)$ matrix,

$$\hat{A}\hat{B} = \hat{C}, \tag{8.20}$$

where, by definition, the elements of $\hat{C}$ are

$$\langle i|\hat{C}|j\rangle = \sum_{k=1}^{n} \langle i|\hat{A}|k\rangle \langle k|\hat{B}|j\rangle. \tag{8.21}$$

It is clear that in general, as for differential operators (see (2.13) and (2.14))

$$[\hat{A}, \hat{B}] \equiv \hat{A}\hat{B} - \hat{B}\hat{A} \neq 0. \tag{8.22}$$

The unit matrix, $\hat{1}$, has all its elements zero except on the diagonal, where its elements are one.

$$\hat{1} = \begin{pmatrix} 1 & 0 \\ 0 & 1 \end{pmatrix}. \tag{8.23}$$

It is again simple to check, using (8.21), that for any matrix $\hat{A}$,

$$\hat{A}\hat{1} = \hat{1}\hat{A} = \hat{A}. \tag{8.24}$$

This is an example of a matrix operator equation analogous to (2.17), which applies directly to the operators, and does not depend on the vector on which they operate.

To each vector $|\psi\rangle$ there corresponds an adjoint vector $\langle\psi|$, with components

$$\langle\psi|i\rangle \equiv \langle i|\psi\rangle^*. \tag{8.25}$$

The scalar product of two vectors is

$$\langle\phi|\psi\rangle \equiv \sum_{i=1}^{n} \langle\phi|i\rangle \langle i|\psi\rangle. \tag{8.26}$$

Two vectors are orthogonal if

$$\langle\phi|\psi\rangle = 0. \tag{8.27}$$

A vector is said to be normalized if

$$\langle\psi|\psi\rangle \equiv \sum_{i=1}^{n} \langle\psi|i\rangle \langle i|\psi\rangle \equiv \sum_{i=1}^{n} |\langle i|\psi\rangle|^2 = 1. \tag{8.28}$$

It is clear that matrix operators have all the general properties of the differential operators discussed in Chapter 2, and can be used to describe operations of observation in the manner developed in Chapter 3. In particular § 3.2 can be applied almost without change except that, for example, (3.1) should now read

$$\bar{a}_\psi = \frac{\sum\limits_{i=1}^{n}\sum\limits_{j=1}^{n}\langle\psi|i\rangle\langle i|\hat{A}|j\rangle\langle j|\psi\rangle}{\sum\limits_{i=1}^{n}\langle\psi|i\rangle\langle i|\psi\rangle}. \tag{8.29}$$

This is an obvious generalization, particularly in view of (8.25).

§ 8.3 Spin

If we take the definition of the angular momentum operators given in § 6.1 and derive the commutation relations which are implied by

$$[\hat{x},\hat{p}_x] = i\hbar, \qquad [\hat{x},\hat{p}_y] = [\hat{p}_x,\hat{p}_y] = [\hat{x},\hat{y}] = 0, \text{etc.,} \tag{8.30}$$

then, (Problem **6.1**)

$$[\hat{l}_x,\hat{l}_y] = i\hbar\hat{l}_z, \tag{8.31}$$

with two other relations obtained by cyclic permutations of the suffices x, y, z. Let us now take these commutation relations to be the defining property of angular momentum operators.

Consider the matrices—the Pauli spin matrices—

$$\hat{\sigma}_x = \begin{pmatrix} 0 & 1 \\ 1 & 0 \end{pmatrix}, \qquad \hat{\sigma}_y = \begin{pmatrix} 0 & -i \\ i & 0 \end{pmatrix}, \qquad \hat{\sigma}_z = \begin{pmatrix} 1 & 0 \\ 0 & -1 \end{pmatrix}. \tag{8.32}$$

Using the definitions given in the previous section, one can check that the matrices,

$$\hat{l}_x = \tfrac{1}{2}\hbar\hat{\sigma}_x, \qquad \hat{l}_y = \tfrac{1}{2}\hbar\hat{\sigma}_y, \qquad \hat{l}_z = \tfrac{1}{2}\hbar\hat{\sigma}_z, \tag{8.33}$$

satisfy the commutation relations (8.31)†. Also one can check directly that

$$\hat{l}^2 \equiv \hat{l}_x^2 + \hat{l}_y^2 + \hat{l}_z^2 = \tfrac{1}{2}(\tfrac{1}{2}+1)\hbar^2\begin{pmatrix} 1 & 0 \\ 0 & 1 \end{pmatrix}, \tag{8.34}$$

so that

$$[\hat{l}_x,\hat{l}^2] = [\hat{l}_y,\hat{l}^2] = [\hat{l}_z,\hat{l}^2] = 0. \tag{8.35}$$

According to (8.17) and (8.18), the eigenvalues of $\hat{l}_z$ are

$$\pm\tfrac{1}{2}\hbar, \tag{8.36}$$

with eigenvectors

$$|u_{+1/2}\rangle = \begin{pmatrix} 1 \\ 0 \end{pmatrix}, \qquad |u_{-1/2}\rangle = \begin{pmatrix} 0 \\ 1 \end{pmatrix}. \tag{8.37}$$

† For a more general discussion see § 12.5(b).

The same vectors are also eigenvectors of $\hat{l}^2$, each corresponding to the eigenvalue

$$\hbar^2 \tfrac{1}{2}(\tfrac{1}{2}+1). \tag{8.38}$$

These are precisely the relations (6.36) and (6.37), which were established for angular momentum quantum numbers in Chapter 6, except that in discussing orbital angular momentum the possible values of l and m, were restricted to integers. By taking the commutation relations (8.31) as the defining property, and allowing matrix representations of the operators, we have uncovered the possibility of

$$l = \tfrac{1}{2}, \qquad m = \pm\tfrac{1}{2}. \tag{8.39}$$

Since this is not associated with orbital motion, it must be the angular momentum—or spin—of the particles themselves. In particular the electron may have spin $\tfrac{1}{2}\hbar$.

In this case, in addition to a factor $\psi(r,\theta,\phi)$ specifying its probability distribution in space, the state function of an electron has a factor

$$|\psi\rangle = \begin{pmatrix} a \\ b \end{pmatrix}, \tag{8.40}$$

specifying its spin. The probability of its having $l_z = \pm\tfrac{1}{2}\hbar$ is given by the generalization of the overlap integral, (3.39),

$$\mathscr{P}_\psi(+\tfrac{1}{2}) = |\sum_{i=1}^{2} \langle u_{+1/2}|i\rangle\langle i|\psi\rangle|^2, \tag{8.41}$$

$$= |a|^2.$$

Similarly

$$\mathscr{P}_\psi(-\tfrac{1}{2}) = |b|^2. \tag{8.42}$$

The normalizing condition is

$$\langle\psi|\psi\rangle \equiv \sum_{i=1}^{2} \langle\psi|i\rangle\langle i|\psi\rangle = |a|^2 + |b|^2 = 1. \tag{8.43}$$

This ensures that the total probability of one or other of the two possible spin orientations is unity—as it should be.

Two refinements in notation may be introduced at this stage. Instead of $|u_{\pm 1/2}\rangle$ we may write $|\pm\tfrac{1}{2}\rangle$ to denote the eigenvector orresponding to this eigenvalue of the diagonal matrix $\hat{l}_z$ (or $\tfrac{1}{2}\hat{\sigma}_z$) Thus the eigenvalue equation is

$$\tfrac{1}{2}\hat{\sigma}_z|\pm\tfrac{1}{2}\rangle = \pm\tfrac{1}{2}|\pm\tfrac{1}{2}\rangle. \tag{8.44}$$

Further we may use $\pm\frac{1}{2}$ as labels to denote the rows or columns in which they appear as eigenvalues. Thus

$$\langle 1|\psi\rangle \equiv \langle +\tfrac{1}{2}|\psi\rangle, \tag{8.45}$$

$$\langle 2|\psi\rangle \equiv \langle -\tfrac{1}{2}|\psi\rangle.$$

In this notation (8.41), which gives the probability that $\hat{l}_z$ has the value $+\frac{1}{2}\hbar$ for the state $|\psi\rangle$, is

$$\mathscr{P}_\psi(+\tfrac{1}{2}) = |\langle +\tfrac{1}{2}|\psi\rangle|^2, \tag{8.46}$$

in close analogy with (3.38).

This notation is quite consistent, but not trivially so. For example the equation

$$\langle +\tfrac{1}{2}|-\tfrac{1}{2}\rangle = 0 \tag{8.47}$$

implies in the old notation, both the vanishing of the first component of $|u_{-1/2}\rangle$,

$$\langle 1|u_{-1/2}\rangle = 0; \tag{8.48}$$

and that its scalar product with $|u_{+1/2}\rangle$ is zero,

$$\langle u_{+1/2}|u_{-1/2}\rangle = 0, \tag{8.49}$$

each of which is correct.

The total z-component of angular momentum of an electron in an atom is now

$$\hat{j}_z = \hat{l}_z + \tfrac{1}{2}\hbar\hat{\sigma}_z \tag{8.50}$$

where $\hat{l}_z$ is the orbital, and $\frac{1}{2}\hbar\hat{\sigma}_z$ the spin contribution. This does not in any way affect the calculation of the energy levels of hydrogen, since the Hamiltonian does not depend on the spin operators. However, there are now two state functions corresponding to the ground state E_1,

$$u_{100}(r)|+\tfrac{1}{2}\rangle, \qquad u_{100}(r)|-\tfrac{1}{2}\rangle. \tag{8.51}$$

The spin provides a basis for the Zeeman splitting of the ground state of hydrogen. To get agreement with experiment the Hamiltonian, (8.7), is now replaced by

$$\hat{H}_\sigma = \hat{H} + \frac{e}{2m_e}B(\hat{l}_z + \hbar\hat{\sigma}_z). \tag{8.52}$$

The two eigenstates, (8.51), are also eigenstates of the new system, since the spin vector factors $|\pm\frac{1}{2}\rangle$ are eigenstates of the operator $\hat{\sigma}_z$. The corresponding levels are

$$E_1^\sigma = E_1 \pm \frac{e\hbar}{2m_e}B, \tag{8.53}$$

7

the extra term arising from the operation of

$$\frac{e\hbar}{2m_e} B\hat{\sigma}_z$$

on the spin eigenvectors $|\pm\frac{1}{2}\rangle$. This gives the required splitting of the ground state into two levels, with a separation

$$\Delta E = \frac{e\hbar B}{m_e}. \tag{8.54}$$

This is in agreement with experiment. Note one curious feature. In order to get quantitative agreement, we did not replace $\hat{l}_z$ in (8.7) by $\hat{j}_z$ (the total z-component of angular momentum) given in (8.50). The factor of $\frac{1}{2}$ in the spin term is missing in (8.52). This is known as the "magnetic anomaly of the spin". We may also say that the magnetic moment of an electron is

$$\mu_e = \frac{e\hbar}{2m_e}, \quad \text{(the Bohr magneton)}, \tag{8.55}$$

since this is the factor by which a field B has to be multiplied to obtain the observed change in energy.

§ 8.4 Statistics and the Exclusion Principle

In the classical mechanics of two identical objects, such as two billiard balls, it is always assumed that the objects can be labelled in some way so that "ball-1 here and ball-2 there" is a distinguishable physical situation from "ball-2 here and ball-1 there". In quantum mechanics it is assumed, as an additional postulate, that for identical quantum particles, such as electrons, no such distinction is possible. Since an energy measurement cannot distinguish between the particles, the Hamiltonian must be symmetric for exchange of particles. Similarly, a state function for two electrons can determine the probability for finding an electron here and another there, but can make no distinction as to which electron is which. The probability distribution is thus also unchanged by exchange of the particles. If x_1 are the general co-ordinates of one particle (which may include specification of spin), and x_2 are the co-ordinates for the other, this postulate then implies that the state function satisfies the relation

$$|\Psi(x_1, x_2)|^2 = |\Psi(x_2, x_1)|^2, \tag{8.56}$$

and hence that

$$\Psi(x_1, x_2) = \pm\, \Psi(x_2, x_1). \tag{8.57}$$

It is found that the choice of positive, or negative sign is a property of the type of particle. For electrons the sign must always be negative. The state function for two electrons is always anti-symmetric for exchange of their co-ordinates. Electrons are then said to satisfy Fermi–Dirac statistics. On the other hand a state function for two photons must always have the positive sign. That is to say, it is always symmetric for exchange. Photons are said to satisfy Einstein–Bose statistics.

For many particle states more complicated symmetries are theoretically possible, but in practice it has been found that the allowed state functions are always either symmetric or anti-symmetric for the exchange of any pair of particles.

Suppose the state function for a system of electrons can be built up as a product of single particle state functions. Since the Hamiltonian is symmetric, the set of possible state functions $\psi_\alpha(x)$, $\psi_\beta(x)$, ... must be the same for all the particles. For two electrons the state function must be

$$\Psi(x_1, x_2) = (\tfrac{1}{2})^{1/2}[\psi_\alpha(x_1)\,\psi_\beta(x_2) - \psi_\beta(x_1)\,\psi_\alpha(x_2)]. \qquad (8.58)$$

If there are n electrons it has the form of a determinant

$$\Psi(x_1, x_2, \ldots, x_n) = \left(\frac{1}{n!}\right)^{1/2} \begin{vmatrix} \psi_\alpha(x_1) & \psi_\beta(x_1) \ldots & \psi_\gamma(x_1) \\ \psi_\alpha(x_2) & \psi_\beta(x_2) \ldots & \psi_\gamma(x_2) \\ \vdots & & \\ \psi_\alpha(x_n) & \psi_\beta(x_n) \ldots & \psi_\gamma(x_n) \end{vmatrix}. \qquad (8.59)$$

Both of these state functions have the required property of anti-symmetry, since exchanging any pair of particle co-ordinates is equivalent to interchanging two rows of the determinant, which changes its sign.

A very important consequence of this is that no two electrons can occupy the same state ($\psi_\alpha \neq \psi_\beta$), since in that case the state function, Ψ, is zero. This is obvious for (8.58). It is also true for the more general state, (8.59), since if two states are the same, two columns of the determinant are the same, and it vanishes identically. This rule, that no two electrons can be in the same state, is known as the *Pauli exclusion principle*.

For photons, or any particles which satisfy Bose statistics and hence require symmetric state functions, we have the same combination of factors as in (8.59), but with all signs positive. In this case any number of particles may be in the same state.

In the classical situation, one can think of the different ψ_α, as specifying possible orbits of the particles. In this case each term in the above expressions represents a classically distinguishable situation. This way of counting the number of different states of the system is known as classical, or Boltzman, statistics.

These different ways of counting the distinguishable states of many particle systems, have very important consequences in statistical mechanics. These can be illustrated by considering the simplest possible system, consisting of two particles, 1 and 2, with two possible states ψ_α and ψ_β. The distinguishable situations are:

Classical statistics:

$$\psi_\alpha(1)\ \psi_\alpha(2),$$
$$\psi_\alpha(1)\ \psi_\beta(2),$$
$$\psi_\beta(1)\ \psi_\alpha(2), \qquad\qquad \text{(4 states)}.$$
$$\psi_\beta(1)\ \psi_\beta(2)$$

Fermi statistics:

$$\psi_\alpha(1)\ \psi_\beta(2) - \psi_\beta(1)\ \psi_\alpha(2) \qquad\qquad \text{(1 state)}.$$

Bose statistics:

$$\psi_\alpha(1)\ \psi_\alpha(2),$$
$$\psi_\alpha(1)\ \psi_\beta(2) + \psi_\beta(1)\ \psi_\alpha(2), \qquad\qquad \text{(3 states)}.$$
$$\psi_\beta(1)\ \psi_\beta(2)$$

At high temperatures all possible states for each type of statistics are equally likely. For Fermi statistics there is no possibility of the particles being in the same state (exclusion principle). For Bose statistics the probability of both particles in the same state is 2/3, for classical statistics it is 1/2. Thus, whereas Fermi statistics prevent particles from getting into the same state and, in this loose sense, keep them apart, Bose statistics increase this probability compared to the classical situation, and so tend to keep them together.

§ 8.5 Atomic Structure

One of the most important consequences of the exclusion principle is that, combined with the level structure of the hydrogen atom, and the notion of electron spin, it explains the periodic table of the elements, which is the basis of the whole of chemistry. If we consider the atomic structure of an atom of charge Z, it will, of course, contain Z electrons. According to the exclusion principle no two of these can be in the same state. However, for a given spatial state, u_{nlm}, two different spin orientations are possible $|+1/2\rangle$ and $|-1/2\rangle$, so that two electrons may be found in each state specified by n, l and m. In the course of time the states of lowest energy, consistent with these restrictions, will get filled up, since electrons in excited states can give off energy by radiation and drop into vacant lower levels. The structure of the periodic table arises because the elements with just sufficient electrons to fill up all the available states, for a given value of n, form closed shells of charge, and are chemically inert systems— the inert gases. The facility for binding to other atoms, which determines the chemical properties of an element, is directly related to the number of electrons extra, or missing, from a closed shell. Thus the closed-shell-plus-one elements are the alkali metals—lithium, sodium, etc. These are very reactive and combine particularly easily with the closed-shell-minus-one elements, the halogens, to form, for example, lithium chloride.

The highly idealized picture outlined above is complicated by the Coulomb repulsions between the different electrons, which we have completely neglected. The detailed theory of atomic structure and quantum physical chemistry are two very large fields, which we will not consider any further. Neither of them involve any new fundamental ideas or basic principles, beyond those introduced above.

§ 8.6 Outline of Further Developments

All the further developments in the theory of atomic structure, which have added to our fundamental understanding, have arisen from the more detailed study of the levels of the simplest atom, hydrogen. We can only sketch these here.

In Part I, and particularly in Chapter 3, we developed techniques for constructing a quantum mechanical theory from a known classical theory. The classical variables become quantum operators, specified by their commutation relations. This procedure has been applied to mechanical systems. It may also be applied to Maxwell's theory of radiation, mentioned in § 1.1. The quantum theory of radiation, so derived, precisely confirms Planck's hypothesis, that the energy of the radiation may be regarded as that of a collection of particles of zero mass (photons), with energies and frequency related according to (1.9).

In interaction with electrons, these photons are emitted or absorbed. The probability for emission, which is a measure of the strength of the interaction, is proportional to the fine structure constant (see § 14.3),

$$\alpha \equiv \frac{e_M^2}{\hbar c} = \frac{1}{137} \tag{8.60}$$

—the dimensionless quantity which can be constructed from the basic physical constants e, $\hbar$ and c. Since Maxwell's theory satisfies the requirements of Special Relativity, the above theory of radiation is a relativistic quantum theory.

The next step is to modify the Schrödinger equation for the hydrogen atom, so that this too is consistent with the theory of Special Relativity. This was done by Dirac† in 1928. The Dirac equation is one of the most remarkable discoveries in physics. It shows that the requirement of Relativity imposed on the quantum theory of the hydrogen atom has the following consequences:

(i) The electron has intrinsic spin $\frac{1}{2}\hbar$.

(ii) The interaction energy with an external magnetic field is as postulated in (8.52) (thus explaining the magnetic anomaly of the spin).

(iii) There are "fine structure" corrections to the Bohr formula for the hydrogen levels. Each level previously specified by n splits into n different levels. These are denoted by

$$E_{nj} = E_n\left[1 + \frac{\alpha^2}{n}\left(\frac{1}{j+\frac{1}{2}} - \frac{3}{4n}\right)\right], \qquad (8.61)$$

where the total angular momentum, j, can take the values

$$j + \tfrac{1}{2} = 0, 1, 2 \ldots n, \qquad (8.62)$$

and the orbital angular momentum is

$$l = j \pm \tfrac{1}{2}. \qquad (8.63)$$

(iv) There is a positively charged counterpart, or anti-particle, to the electron—the positron.

Results (i) and (ii) were known previously, but had been fed into the non-relativistic theory, simply to obtain agreement with experiment. The Dirac equation shows that both effects are required for fundamental reasons. The fine structure formula, (8.61), for the energy levels is in agreement with experiment. Remarkably, it had actually been derived in 1916 by Sommerfeld on the basis of Old Quantum Theory (but with an incorrect interpretation of the quantum numbers). The result (iv) was a prediction, which has since been amply confirmed by experiment.

In both the Dirac and Schrödinger equations for hydrogen, the only electromagnetic effect included is the Coulomb potential between the

† P. A. M. Dirac, *Proc. Roy. Soc.* **A117**, 610 (1928).

proton and the electron. But the electron accelerates in its orbit, and is thus a source of radiation, which may react back on the electron itself. This produces a further splitting of levels, which are degenerate according to the fine structure formula, (8.61). This type of splitting was measured first by Lamb in 1947 and is known as the Lamb shift. The corresponding calculation involves a very complicated interplay of the Dirac theory of the electron and the quantum theory of radiation. The theoretical value is in complete agreement with experiment.

A very similar effect is found for the motion of an electron in a magnetic field. Here again the electron's radiation can react back on the electron, and produce a correction to the electrons magnetic moment as stated in (8.55). These corrections have been calculated to order α^6, and are in complete agreement with experiment. There are, at the time of writing, no known discrepancies relating to the interaction of radiation (photons) with electrons, that indicate any fundamental limitations to the present theory of relativistic quantum electrodynamics.

A very high degree of accuracy is involved in these results. We may express the various corrections to the Bohr formula as fractions of the hydrogen ground state binding energy. The correction for proton recoil is (see § 7.5),

$$\frac{\Delta E_{\text{recoil}}}{|E_1|} \simeq \frac{m_e}{m_p} \simeq 10^{-3}.$$

The fine structure corrections implied by (8.61), are of order

$$\frac{\Delta E_{fs}}{|E_1|} \simeq \alpha^2 \simeq 10^{-4}.$$

The Lamb shift correction is

$$\frac{\Delta E_L}{|E_1|} \simeq \alpha^3 \simeq 10^{-6}.$$

Specifically, the binding energy of the ground state of hydrogen is

$$|E_1| = 13 \cdot 605 \text{ eV}.$$

In terms of frequency this is one Rydberg

$$R_\infty = \frac{|E_1|}{2\pi\hbar} \quad \begin{aligned} &= 3 \cdot 28985 \times 10^{15} \text{ cycles/sec.} \\ &= 3 \cdot 28985 \times 10^9 \text{ Mc/sec.} \end{aligned}$$

The fine structure splitting between the $2P_{1/2}$ $(n = 2, l = 1, j = \frac{1}{2})$ and the $2P_{3/2}$ levels (which are degenerate according to (8.61)), is

$$\Delta E_{fs}(2P_{1/2} - 2P_{3/2}) = 1 \cdot 10 \times 10^4 \text{ Mc/sec.}$$

The Lamb shift between the $2P_{1/2}$ and $2S_{1/2}$ levels, which are degenerate according to (8.61), is

$$\Delta E_{\text{Lamb}}(2S_{1/2} - 2P_{1/2}) = 1 \cdot 057 \times 10^3 \text{ Mc/sec.}$$

Since this figure is accurate to one megacycle, the measurement and theoretical calculation are in agreement to one part in 10^{-9} of the ground state binding energy.

PROBLEMS VIII

8.1. Show that the Pauli spin matrices, (8.32), satisfy the relations

$$\hat{\sigma}_x \hat{\sigma}_y = -\hat{\sigma}_y \hat{\sigma}_x = i\hat{\sigma}_z,$$

$$\hat{\sigma}_y \hat{\sigma}_z = -\hat{\sigma}_z \hat{\sigma}_y = i\hat{\sigma}_x,$$

$$\hat{\sigma}_z \hat{\sigma}_x = -\hat{\sigma}_x \hat{\sigma}_z = i\hat{\sigma}_y.$$

8.2. The spin and angular factor in the state function of an electron is

$$\psi(\theta, \phi)|\chi\rangle = \frac{1}{\sqrt{3}} Y_1^0 |+\tfrac{1}{2}\rangle + \sqrt{\left(\frac{2}{3}\right)} Y_1^1 |-\tfrac{1}{2}\rangle.$$

Show that the angular distribution, after summing over both spin directions, is isotropic.

PART III

Nuclear Physics

CHAPTER 9

RUTHERFORD SCATTERING AND
α-DECAY

§ 9.1 Rutherford Scattering

We now turn to the consideration of the nuclei of the atoms. The first thing to determine is their size, and this was done in the classic experiment of Rutherford in 1912.

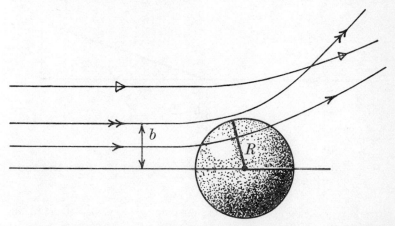

FIG. 9.1. The diagram shows the deflection produced by an extended charge distribution, of radius R, for trajectories with different impact parameters, b. The greatest deflection is for a trajectory which grazes the edge and, hence, for which $R \simeq b$.

The experiment consisted in studying the deflections produced in the paths of α-particles in their passage through thin gold foil. The largest deflections were assumed to be due to the Coulomb repulsion between the α-particles (charge $Z_\alpha = 2$), and the gold nuclei (charge $Z = 79$). Suppose the gold nucleus has radius R. Figure 9.1 shows three typical α-particle trajectories, which may be specified by their impact parameters, b. Clearly for $b > R$, the smaller b, the closer the α-particle approaches the nucleus, the stronger the repulsion, and hence the larger the deflection. However, for $b < R$, the α-particle starts to plough through the middle of the charge distribution. The

forces producing the deflection are strongest while the particle is
inside the nucleus, but now part of the charge is on the "wrong" side
of the trajectory, and works in the opposite direction to the bulk of
the nucleus. The nucleus is thus less effective in producing a deflection.
The maximum deflection occurs when the particle just grazes the edge
of the nucleus, and the radius of the charge distribution may be
approximately equated to the impact parameter corresponding to the
maximum deflection.

This argument can be made more quantitative using purely
classical notions. Consider a trajectory with small deflection. Along
the original line of flight the particle is slowed down on its way in, and
speeded up on its way out. There is no net effect, so the longitudinal
forces may be neglected. The main deflection is produced just as the

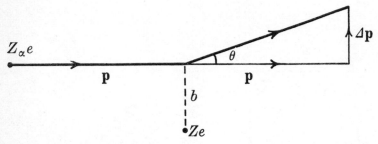

FIG. 9.2. Sketch of a particle trajectory with impact parameter b and
(small) deflection θ.

particle passes the nucleus. We approximate the Coulomb repulsion
in this region by a force of

$$F = \frac{Z_\alpha Z e_M^2}{b^2} \tag{9.1}$$

operating perpendicular to the original direction, over a distance b.
If the particle has velocity v, the force operates for a time $\Delta t = b/v$.
This will produce a transverse momentum according to Newton's
Law, $(\Delta p/\Delta t = F)$,

$$\Delta p = F \Delta t$$
$$= \frac{Z_\alpha Z e_M^2}{b^2} \frac{b}{v}. \tag{9.2}$$

The resulting deflection is (see Fig. 9.2),

$$\theta \simeq \frac{\Delta p}{p} = \left(\frac{Z_\alpha Z e_M^2}{bv}\right)\bigg/(mv), \tag{9.3}$$

where m is the α-particle mass. Thus the approximate relation between deflection and impact parameter is

$$b = \frac{Z_\alpha Z e_M^2}{mv^2 \theta}. \tag{9.4}$$

The maximum deflections observed by Rutherford in the α-particle experiment were of the order of 1 radian and, as explained above, the corresponding impact parameter may be approximately identified with the radius. Thus, ($\theta = 1, b = R$ in (9.4)),

$$R \simeq \frac{Z_\alpha Z e_M^2}{mv^2}. \tag{9.5}$$

Taking

$$m = 6 \times 10^{-27} \text{ kg}, \tag{9.6}$$

and a velocity somewhat less than the velocity of light,

$$v \simeq 10^7 \text{ m/sec}, \tag{9.7}$$

we get

$$R \simeq 10^{-14} \text{ m} \tag{9.8}$$

—a remarkably small distance even on the atomic scale, differing by a factor of 10^4 from the Bohr radius, (1.24).

§ 9.2 Nuclear Interactions

The charge on a nucleus is carried by protons, each of which has the same charge as an electron, but of course of the opposite sign. The mass of a proton is

$$m_p = 1 \cdot 6 \times 10^{-27} \text{ kg},$$
$$= 1836 m_e,$$
$$= 938 \text{ MeV/c}^2. \tag{9.9}$$

The mass of a nucleus is, in general, roughly double that of the total mass of the protons required to make up its total charge. The surplus mass is made up of neutrons, which are neutral particles of approximately the same mass as a proton ($m_n = 939 \text{ MeV/c}^2$). We use the term *nucleon* to denote either a neutron or proton.

From the point of view of chemistry and atomic physics, the nuclei are extremely stable objects, remaining unchanged by the most violent chemical reactions. The question immediately arises, what are the forces which hold these extremely stable structures together.

The only forces known to classical and atomic physics are gravitational and electromagnetic (particularly the Coulomb force between charges). Since the distances involved in a nucleus are smaller by a factor of 10^4 than those in an atom, the Coulomb *repulsion* between protons in a nucleus is larger than the attraction between the nucleus and the electrons in an atom by 10^8. This force is tending to blow the nucleus apart. The gravitational force is attractive, but the ratio of gravitational attraction to Coulomb repulsion between two protons at any arbitrary distance apart is

$$\frac{F_g}{F_e} = \frac{\gamma m_p^2}{e_M^2}, \tag{9.10}$$

where γ is the gravitational constant. Since γ is $10^{-11} \times 6 \cdot 6$ mks, this ratio is 10^{-36}, and the gravitational force is completely negligible inside the nucleus—as indeed it is in atomic structure and in the chemical reactions between atoms, as assumed in Part II.

The simplest conclusion is that inside the nucleus, over distances of about 10^{-14} m, there is a specifically nuclear interaction operating between nucleons, which is strong enough to overcome the enormous Coulomb repulsions between the tightly packed protons. This force is evidently more than a million times stronger than anything encountered in chemistry. Since it plays no role in the electronic structure of an atom, it must be a short range force, effective only over distances of about 10^{-14} m. The first problem of nuclear physics is to obtain a detailed understanding of it, in the same way that we think we understand the gravitational interaction between masses, and the Coulomb interaction between charged particles.

§ 9.3 α-Decay

Before considering the problem of nuclear interactions, it is important to check that quantum mechanics, developed above for atomic systems, remains valid without fundamental modifications inside the nucleus. Strong evidence that this is so comes from the theory of α-decay. In this process a (parent) nucleus, A, decays spontaneously into an α-particle and a (daughter) nucleus D.

$$A \rightarrow \alpha + D. \tag{9.11}$$

The α-particle is a helium nucleus and is actually made up of 2 protons and 2 neutrons. In discussing the α-decay of a heavy parent nucleus, say radium, we can consider the α-particle as a single particle, and regard the rest of the nucleus, the daughter, as fixed. (See the discussion at the end of § 7.5) We have talked above of nuclear forces, but nuclei are certainly not classical systems, and we have seen the notion of force plays no direct role in quantum mechanics. We must consider rather the interaction potential. At distances large compared with 10^{-14} m the only interaction is the repulsive Coulomb potential between the α-particle (charge Z_α) and the daughter nucleus (charge Z). However, at distances less than

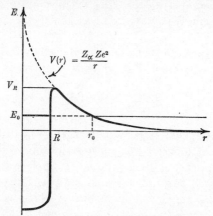

FIG. 9.3. Energy diagram of the mutual potential between an α-particle and "daughter" nucleus. For large distances ($r \gg 10^{-14}$ m) it is simply the Coulomb repulsion. At short distances it is dominated by the nuclear attraction.

10^{-12} cm the strongly attractive nuclear potential must dominate, and the net effect must be something of the general form illustrated in Fig. 9.3. It is convenient to define the energy

$$V_R = \frac{Z_\alpha Z e_M^2}{R},\qquad(9.12)$$

which, from Fig. 9.3, is a reasonable estimate of the maximum height of the potential between the α-particle and daughter nucleus.

The qualitative explanation of α-decay has already been anticipated in Chapter 4. Classically, if a particle is situated inside such a potential well, with energy E_0 satisfying

$$V_R > E_0 > 0,\qquad(9.13)$$

it is bound, and there is no possibility of its escape. What is observed, however, is that each radio-active nucleus shows a constant probability per unit time, $1/\tau$, that it will decay. Thus, if $N(t)$ is the number of nuclei at time t,

$$\frac{dN(t)}{dt} = \frac{-N(t)}{\tau},\qquad(9.14)$$

so that

$$N(t) = N(0) e^{-t/\tau}. \tag{9.15}$$

(The constant τ is called the mean life of the particular nucleus.) Quantum mechanically the possibility of barrier penetration enables the α-particle to escape, and provides the mechanism for the observed decay of the system. We must now make this picture more precise and quantitative.

We have seen that the typical nuclear length is R. Thus the typical nuclear time, on dimensional grounds, is

$$\tau_n = \frac{m_p R^2}{\hbar} \simeq 10^{-21} \text{ sec.} \tag{9.16}$$

Since the observed mean lives vary from 10^{-7} sec to 10^{10} years, even the fastest α-decay proceeds very very slowly on a nuclear time scale. To a good first approximation the daughter nucleus and the α-particle may be considered to form a stable system. For simplicity let us suppose that the angular momentum involved is zero. Then, in principle, the energy E_0 is obtained by solving the eigenvalue problem for the bound system in which the potential $V^1(r)$ is similar to that drawn in Fig. 9.3 for $r < R$, but for larger values it is assumed to remain constant.

$$V^1(r) = V_R, \qquad r \geqslant R. \tag{9.17}$$

This would require the solution for E_0 of the eigenvalue equation, (7.10), for $l = 0$,

$$\left[\frac{-\hbar^2}{2\mu} \frac{\partial^2}{\partial r^2} + V^1(r) - E_0 \right] \chi(r) = 0, \tag{9.18}$$

where μ is the reduced mass of the α-particle and daughter nucleus. However, we do not know $V^1(r)$ in any detail, and for our present purpose it is sufficient to take E_0 directly from experiment. Since we have taken as our zero of energy the energy of the system when the daughter nucleus D and the α-particle are infinitely separated, E_0 is simply the difference between the rest energy of the parent nucleus, A, and that of the daughter nucleus and the α-particle.

$$E_0 = m_A c^2 - (m_D + m_\alpha) c^2. \tag{9.19}$$

If we consider the realistic problem, we have to solve

$$\left[\frac{-\hbar^2}{2\mu} \frac{\partial^2}{\partial r^2} + V(r) - E_0 \right] \chi(r) = 0 \tag{9.20}$$

for $\chi(r)$, with $V(r)$, now representing the actual potential. Define

$$K^2(r) = \frac{2\mu}{\hbar^2}(V(r) - E_0), \qquad (9.21)$$

then

$$\frac{\partial^2 \chi(r)}{\partial r^2} - K^2(r)\,\chi(r) = 0. \qquad (9.22)$$

Note that although this is the full three-dimensional problem, the equation to be solved is very similar to (4.26) and (4.27) discussed in Chapter 4. If we look for a solution of the form

$$\chi(r) = a\,e^{-f(r)}, \qquad (9.23)$$

substituting (9.23) into (9.22) gives

$$-\frac{d^2 f}{dr^2} + \left(\frac{df}{dr}\right)^2 - K^2 = 0. \qquad (9.24)$$

For values of r for which the particle is penetrating the barrier, we may expect that f is a slowly varying function (see Fig. 4.2), so that

$$\frac{d^2 f}{dr^2} \simeq 0,$$

and hence the approximate solution to (9.24) is

$$f(r) = \int^r K(r')\,dr'. \qquad (9.25)$$

As can be seen from Fig. 9.3, the α-particle emerges from the potential barrier at $r = r_0$, where

$$\frac{Z_\alpha Z e_M^2}{r_0} = E_0. \qquad (9.26)$$

According to (9.23) and (9.25), the reduction in the amplitude of the state function in passing through the barrier is

$$\frac{\chi(r_0)}{\chi(R)} = \exp\left[-\int_R^{r_0} K(r)\,dr\right]. \qquad (9.27)$$

8

The square of this quantity is the relative probability of finding the particle at $r = r_0$ and at $r = R$. This is the transmission coefficient, T, which may be interpreted semi-classically as the probability of the particle penetrating the barrier whenever it hits it (see (4.35)). Thus

$$T = \exp\left[-2 \int_R^{r_0} \left(\frac{2\mu}{\hbar^2} (V(r) - E_0) \right)^{1/2} dr \right]. \qquad (9.28)$$

The probability per unit time for escape is T multiplied by the frequency of oscillation of the α-particle within the potential well, which may be taken to be the inverse of the typical nuclear time. Thus

$$\frac{1}{\tau} = \frac{T}{\tau_n}. \qquad (9.29)$$

The integral (9.28) can be evaluated without much difficulty, but to understand the general features of the system, we may take the integrand to be constant at its average value. Then

$$T = \exp\left[-2 \left(\frac{2\mu}{\hbar^2} \right)^{1/2} \left(\frac{V_R - E_0}{2} \right)^{1/2} (r_0 - R) \right], \qquad (9.30)$$

which is just the expression (4.35). We may substitute for r_0 and R from (9.26) and (9.12). Also, since the nuclei which decay by α-emission all have mass greater than 200 in proton units, the reduced mass is essentially that of the α-particle itself. Thus

$$\mu \simeq m_\alpha \simeq 4m_p, \qquad (9.31)$$

and, eliminating r_0 by (9.26),

$$T = \exp\left[\frac{-4(V_R - E_0)^{3/2}}{E_0 \hbar / (m_p^{1/2} R)} \right]. \qquad (9.32)$$

The quantity $\hbar / (m_p^{1/2} R)$ has the dimensions of (energy)$^{1/2}$, and is approximately unity when expressed in terms of (MeV)$^{1/2}$ for $R = 10^{-14}$ m. Hence the mean life expressed only in terms of the energy available for the reaction is

$$\frac{1}{\tau} = \frac{1}{\tau_n} \exp\left[-4 \frac{(V_R - E_0)^{3/2}}{E_0} \right]$$
$$= 10^{21} \exp\left[-4 \frac{(25 - E_0)^{3/2}}{E_0} \right] \text{sec}^{-1}. \qquad (9.33)$$

where V_R and E_0 are expressed in MeV and, in the final expression V_R has been evaluated for a typical value of $Z = 80$. The important point about this formula is its very strong dependence on E_0, the available energy, which accounts for the enormous variation in observed mean life times for essentially the same nuclear radius for heavier nuclei. Thus if

$$E_0 = 4 \text{ MeV}, \qquad \tau \simeq 3 \times 10^{12} \text{ years}, \qquad (9.34)$$

$$E_0 = 8 \text{ MeV}, \qquad \tau \simeq 10^{-6} \text{ sec},$$

while experimentally, for example,

$$E_0 = 4 \cdot 3 \text{ MeV} \qquad \tau = 2 \times 10^{10} \text{ years} \quad \text{(Thorium)}, \qquad (9.35)$$

$$E_0 = 7 \cdot 83 \text{ MeV} \qquad \tau = 10^{-3} \text{ sec} \quad \text{(Radium } C'),$$

The rough theory presented here is in reasonable agreement with experiment considering that we have made rather crude assumptions in evaluating the exponent.

If the expression (9.28) is evaluated more accurately, the argument can be turned round, and the observed lifetimes, τ, and released energies, E_0, can be used to determine more precisely the radii of the α-active nuclei. These calculations show that nucleons in a nucleus pack tight, like billiard balls in a string bag, each one taking up a spherical volume of radius $1 \cdot 5 \times 10^{-15}$ m, so that the radius of a nucleus with A nucleons is

$$R = 1 \cdot 5 \times 10^{-15} A^{1/3} \text{m}. \qquad (9.36)$$

This result is completely consistent with the conclusions from the Rutherford experiment (§ 9.1), which are based on absolutely different considerations.

§ 9.4 Summary

We have established that atomic nuclei have radii of the order of 10^{-12} cm. They are made up of neutrons and protons (nucleons), which are held together by a specifically nuclear potential, which is extremely powerful over a range comparable with the nuclear radius, being strong enough to swamp the very strong Coulomb repulsions between the tightly packed protons in a nucleus. The motions of the nucleons in this nuclear potential appear to be governed by quantum mechanics, developed to deal with the mechanics of atoms, since this gives an excellent understanding of α-decay. This essentially nuclear phenomenon is quite unintelligible on the basis of classical mechanics (or Old Quantum Theory). However, the nature of the nuclear potential is quite unknown, and to investigate this is the major problem of nuclear physics.

PROBLEMS IX

9.1. Evaluate the integral in (9.28) and hence show that a more accurate expression for T is

$$T = \exp\left[\frac{-2\sqrt{(2\mu E_0)}}{\hbar} r_0\left\{\frac{\pi}{2} - 2\left(\frac{R}{r_0}\right)^{1/2}\right\}\right].$$

Hint. Make the substitution

$$\frac{E_0 r}{Z_\alpha Z e_M^2} \equiv \frac{r}{r_0} = \cos^2 x.$$

The integral is then straightforward giving

$$T = \exp\left[\frac{-2\sqrt{(2\mu E_0)}}{\hbar} r_0\left\{\cos^{-1}\left(\frac{R}{r_0}\right)^{1/2} - \left(\frac{R}{r_0}\right)^{1/2}\left(1 - \frac{R}{r_0}\right)^{1/2}\right\}\right].$$

This integral can be expanded in powers of R/r_0.

9.2. Use this expression to derive a more accurate estimate of the lifetime of Radium C', ($A = 210$, $Z = 82$, $E_0 = 7\cdot8$ MeV).

CHAPTER 10

SCATTERING THEORY

§ 10.1 Introduction

The only way to investigate the nuclear potential is to bring the nucleons together and study their mutual interactions, much as one studies magnetic forces by bringing two magnets together to find how they react on each other. To do this systematically one needs a beam of nuclear particles; a target of nuclei (or nucleons); and some detection device to see how the beam is deflected—or scattered—by the nuclear interactions between the particles in the beam and those in the target. The typical nuclear experiment is thus like playing a hose on some object, and trying to deduce the shape of the object—analogue of the shape of the interaction potential—from a detailed study of the angular distribution and intensity of the splash. Since the beam is scattered by the target, these are known as scattering experiments, and they are analysed by scattering theory. We are dealing with quantum systems so this has to be a quantum scattering theory, but to clear one's ideas it is very useful to see first how such experiments go under purely classical conditions.

§ 10.2 Classical Scattering Theory

Consider a uniformly dense beam of particles, all travelling in the same direction with velocity v. The flux, F, of the beam is the number of particles crossing unit area (perpendicular to the beam direction) per unit time. This is the number of particles in a volume of unit cross-section and length v. Thus, if the particle density is ρ,

$$F = \rho v. \tag{10.1}$$

The dimensions of flux are

$$[F] = L^{-2} T^{-1}. \tag{10.2}$$

Take the origin of co-ordinates at the target, and the z-axis in the direction of the beam (Fig. 10.1). The intensity and direction of the scattering—or splash—is measured by the, so called, "differential cross-section", $\sigma(\theta, \phi)$, where

$$\sigma(\theta, \phi)\, d\Omega = \text{(number of particles into the solid angle } d\Omega(\theta, \phi)$$

$$\text{per unit time, per unit flux)} = d\sigma(\theta, \phi). \tag{10.3}$$

The dimensions of $\sigma(\theta, \phi)$ are

$$[d\sigma/d\Omega] = [\sigma(\theta, \phi)] = T^{-1} \, (L^{-2} \, T^{-1})^{-1} = L^2, \qquad (10.4)$$

which is an area.

The total cross-section σ is the integral of the differential cross-section over all solid angles,

$$\sigma = \int \sigma(\theta, \phi) \, d\Omega, \qquad (10.5)$$

which is also an area. It is evident from the definitions that the total cross-section is the total number of particles deflected in any direction, per unit flux, per unit time. But the deflected particles are just

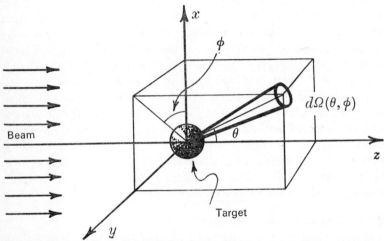

FIG. 10.1. The conventional polar co-ordinates used to describe scattering. The origin is taken at the target and the z—or polar—axis in the direction of the beam.

those which hit the target, and unit flux is one particle per unit area, per unit time. So the total cross-section is the cross-sectional area which the target presents to the direction of the beam. Hence the name!

By the same token, the differential cross-section, $\sigma(\theta, \phi) \, d\Omega$, is the effective area of the target which gives rise to deflections into the solid angle $d\Omega(\theta, \phi)$.

Since nuclear radii are of the order of 10^{-14} m, the target area due to a nuclear target is of the order of the square of this, and nuclear cross-sections are measured in barns, where

$$1 \text{ barn} = 10^{-24} \text{ cm}^2.$$

Consider a classical system which has azimuthal symmetry, so that the deflection produced is independent of ϕ. Let $b(\theta)$—$b(\theta) + \delta b$ be the range of impact parameters for trajectories which suffer deflections

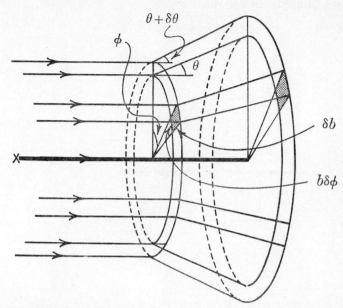

FIG. 10.2. Diagram showing the effective area $b(\theta)\,db\,d\phi$ for scattering into a solid angle $d\Omega(\theta, \phi)$.

through angles varying from θ to $\theta + \delta\theta$. Then, from Fig. 10.2, the effective area for producing a deflection into a solid angle $d\Omega(\theta, \phi)$ is

$$d\sigma(\theta, \phi) = \sigma(\theta, \phi)\,d\Omega(\theta, \phi) = b(\theta)\,db\,d\phi. \tag{10.6}$$

The solid angle is

$$d\Omega = \sin\theta\,d\theta\,d\phi. \tag{10.7}$$

Since $\sigma(\theta, \phi)$ is assumed independent of ϕ, both sides of (10.6) can be integrated over ϕ from 0 to 2π to give

$$2\pi\sigma(\theta)\sin\theta\,d\theta = 2\pi b(\theta)\,db. \tag{10.8}$$

Thus

$$\frac{d\sigma}{d\Omega} = \sigma(\theta) = \frac{b(\theta)}{\sin\theta}\left|\frac{db}{d\theta}\right|. \tag{10.9}$$

(The modulus signs are put on the derivative, since $\sigma(\theta)$ is necessarily positive.)

(a) *Classical Hard Sphere Scattering*

As an extreme example consider scattering from a hard sphere, of radius a, from which the beam particles bounce elastically. It is evident from Fig. 10.3 that

$$b(\theta) = a \sin \frac{\pi - \theta}{2},$$

$$= a \cos(\theta/2). \qquad (10.10)$$

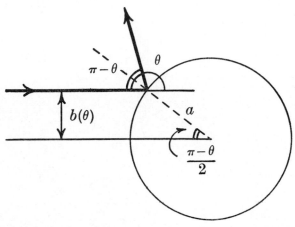

FIG. 10.3. Scattering by a hard sphere of radius a showing the relation between $b(\theta)$ and the scattering angle θ.

Thus

$$\left| \frac{db}{d\theta} \right| = \frac{a}{2} \sin(\theta/2). \qquad (10.11)$$

So that, by (10.9), the differential cross-section is

$$\frac{d\sigma}{d\Omega} = \sigma(\theta) = \frac{a^2}{4}. \qquad (10.12)$$

The total cross-section is

$$\sigma = \int \frac{a^2}{4} d\Omega = \pi a^2. \qquad (10.13)$$

The final result is a simple example of the general statements made above about total cross-sections, since πa^2 is just the total area presented by the hard sphere in the direction of the beam. The total cross-section—the total amount of splash—determines the effective size of the target.

Information on the shape of the target is given by the differential cross-section. In the case of a hard sphere this is isotropic—the same in all directions—at all energies of the beam.

(b) Coulomb Scattering

A completely different scattering distribution is obtained, if the beam consists of particles of charge Z_1, and the deflections are produced by the Coulomb repulsions from the charge Z_2 on the (fixed) target (as in the Rutherford experiment). Using the approximations of Chapter 9, the relation (9.4) between impact parameter and deflection can be expressed in terms of beam momentum,

$$b(\theta) = \frac{Z_1 Z_2 e_M^2 m}{p^2 \theta}. \qquad (10.14)$$

Thus

$$\left| \frac{db}{d\theta} \right| = \frac{Z_1 Z_2 e_M^2 m}{p^2 \theta^2}, \qquad (10.15)$$

and for small θ (so that $\sin \theta \simeq \theta$), according to (10.9),

$$\sigma(\theta) = \left(\frac{Z_1 Z_2 e_M^2 m}{p^2} \right)^2 \frac{1}{\theta^4}. \qquad (10.16)$$

Thus, in contrast to hard sphere scattering, the scattered particles due to Coulomb interaction are very sharply peaked forward. As the momentum, p, increases, θ has to decrease to give the same $\sigma(\theta)$, so the forward peaking becomes more marked as the momentum—or energy—of the beam increases.

To take this over into quantum theory the reaction between beam particles and target must be expressed in terms of the interaction potential (Fig. 10.4). For the hard sphere we have

$$V(r) = 0, \qquad r > a,$$
$$V(r) \to \infty, \qquad r = a. \qquad (10.17)$$

On the other hand, the repulsive Coulomb potential is

$$V(r) = \frac{Z_1 Z_2 e_M^2}{r}. \qquad (10.18)$$

The hard sphere is the extreme form of a "hard" potential. This means that the potential rises very abruptly, which implies the sudden onset of very strong forces, when the particles come within a certain distance of each other. Such potentials lead typically to nearly isotropic scattering at all energies.

By contrast the Coulomb interaction is typical of a very "soft" potential, representing a force which varies slowly with distance and operates weakly over a very wide range of separation between the particles. Such potentials give rise to scattering mainly through small angles, and the forward cone which contains most of the scattered particles tends to narrow in angle as the energy of the beam particles increases.

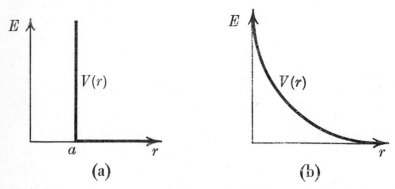

FIG. 10.4. The interaction potentials. (*a*) The hard sphere. (*b*) A repulsive Coulomb interaction.

By studying the dependence on energy and angle of the scattering of nuclear particles, a sharpening of the sort of considerations outlined above should enable one to deduce the shape of the potential operating between them. This is the basic tool of nuclear physics.

§ 10.3 Quantum Scattering Theory

The nuclear interaction operates over distances of about 10^{-12} cm. The beam particles travel with speeds comparable with the velocity of light, say, 10^9 cm/sec. Therefore, the transit times in a nuclear collision—the times taken for beam particles to cross the region of interaction—are of the order of 10^{-21} sec. Since $\hbar$ is about 10^{-27} cgs, the energy of the beam must be *much* greater than 10^{-5} erg—or 10 MeV—for classical notions, such as the impact parameter of a beam trajectory, to have any validity. Nuclear scattering experiments must be analysed in terms of quantum mechanics, the extreme quantum limit being low-energy scattering experiments. (It so happens that the classical theory of the Rutherford experiment outlined in § 9.1 leads to the same conclusions as a more correct quantum treatment along the lines developed below.)

Consider the scattering of a beam of particles by a target fixed at the origin of co-ordinates, when the interaction potential between the beam particles and the target is $V(r)$, and the energy of the beam is

$$E = \frac{p^2}{2m}. \tag{10.19}$$

The state function must be an energy eigenfunction of the corresponding Schrödinger equation for this energy. If the potential has a finite range the solution must be of such a form that, for large distances from the target, it represents a plane beam in the z-direction and a scattered wave, which must consist of outgoing spherical waves only. To save numerous factors of $\hbar$, we introduce

$$k = \frac{p}{\hbar}, \tag{10.20}$$

then asymptotically (for large r)

$$u_k(r) \sim \qquad e^{ikz} \qquad + \qquad f(\theta, \phi) \frac{e^{irk}}{r}, \tag{10.21}$$

$$= \text{(plane wave)} + \text{(scattered wave)}.$$

The particle density in the plane wave is

$$\rho = |e^{ikz}|^2 = 1. \tag{10.22}$$

The velocity is

$$v = \frac{\hbar k}{m}. \tag{10.23}$$

Thus the flux, defined above, (10.1), is

$$F = \rho v = v. \tag{10.24}$$

The number of scattered particles in the volume between r and $r + dr$ in the solid angle $d\Omega(\theta, \phi)$ is

$$\left| \frac{f(\theta, \phi) e^{ikr}}{r} \right|^2 r^2 d\Omega \, dr = |f(\theta, \phi)|^2 d\Omega \, dr. \tag{10.25}$$

The number into $d\Omega$ per unit time is

$$|f(\theta, \phi)|^2 d\Omega v.$$

Thus, the differential cross-section—the number into $d\Omega$, per unit time, per unit flux—is

$$d\sigma = \sigma(\theta, \phi)\, d\Omega = |f(\theta, \phi)|^2 d\Omega. \qquad (10.26)$$

The total cross-section is

$$\sigma = \int |f(\theta, \phi)|^2 d\Omega.$$

Thus, the experiment measures $|f(\theta, \phi)|^2$. This is related to $V(r)$, since (10.21) is the asymptotic form of a solution to the Schrödinger equation with $V(r)$ as potential energy. In nearly all cases of physical interest both $f(\theta, \phi)$, and hence $\sigma(\theta, \phi)$ are, in fact, independent of ϕ. In algebraic terms, the problem posed at the end of the last section is to obtain information about $V(r)$ from a study of $|f(\theta)|^2$.

§ 10.4 Phase Shift Analysis

To develop a classical theory of scattering it is convenient to analyse the beam in terms of the impact parameters of the different trajectories. This is not possible quantum mechanically, since a beam of particles of definite momentum does not have any definite position. However, if the beam momentum is p, and the impact parameter of a particular particle is b, then classically the angular momentum about the origin is

$$pb \simeq \hbar l. \qquad (10.27)$$

This suggests that we analyse the beam in terms of its angular momentum components. If the radius of the interaction is R, there will only be scattering if the particle hits the target, which requires

$$b \leqslant R$$

or, by (10.20) and (10.27),

$$l < kR. \qquad (10.28)$$

If the energy of the beam is sufficiently small, so that

$$kR < 1, \qquad (10.29)$$

there will only be scattering for $l = 0$, (S-wave). We have obtained the condition (10.29) by classical considerations. A similar result can be derived from quantum mechanics. Equation (10.29) specifies the extreme quantum limit, which we now investigate in some detail.

Given a plane beam in the z-direction, of momentum $p \, (= \hbar k)$, the state function is

$$u_k(r, \theta, \phi) = e^{ikz} = e^{ikr\cos\theta}. \qquad (10.30)$$

This includes all components of angular momentum about the origin. The component with angular momentum $l = 0$ is given by the overlap of (10.30) with the appropriate eigenfunction, $Y_0^0(\theta, \phi)$. Thus

$$u_{k,S}(r) = \frac{1}{\sqrt{4\pi}} \int u_k(r, \theta, \phi)\, Y_0^0(\theta, \phi)\, d\Omega$$

$$= \frac{1}{4\pi} \int\limits_0^{2\pi} d\phi \int\limits_0^{\pi} e^{ikr\cos\theta} \sin\theta\, d\theta.$$

The normalization has been chosen so that a state function $u(r)$, which is a function of r only, and thus pure S-wave, is unchanged by the operation. The integral over ϕ is trivial. The integral of θ can be performed by the substitution

$$\cos\theta = w,$$

$$-\sin\theta\, d\theta = dw,$$

$$u_{k,S}(r) = \tfrac{1}{2} \int\limits_{-1}^{+1} e^{ikrw}\, dw$$

$$= \frac{e^{ikr} - e^{-ikr}}{2ikr}. \tag{10.31}$$

Notice that this is a combination of outgoing and incoming *spherical* waves. This is an essentially quantum mechanical picture, and our discussion at this point is completely different from the classical particle picture.

If we now pick out the S-wave part of the typical scattering state function, (10.21), the required part of the beam is (10.31). The S-wave part of the scattered wave is that part for which $f(\theta)$ is a constant, since any function of r only corresponds to $l = 0$. So the asymptotic form of the S-wave part of the state function is

$$u_S(r) \sim \frac{e^{ikr} - e^{-ikr}}{2ikr} \quad + \quad f\frac{e^{ikr}}{r}, \tag{10.32}$$

$$= (l = 0 \text{ part of beam}) + (l = 0 \text{ part of scattering}).$$

The S-wave scattering at some fixed energy is thus determined by a number, f, which may be complex. We now show that the scattering can in fact be expressed in terms of a single real number.

If there is no scattering potential the S-wave part of the plane beam is given by (10.31). The effect of the scattering potential can only be to alter the *outgoing* wave, since the scattering consists purely of outgoing waves. It is thus of the form

$$u_S(r) \sim \frac{S\,e^{ikr} - e^{-ikr}}{2ikr}. \tag{10.33}$$

The flux of the incoming spherical wave is proportional to

$$|e^{-ikr}|^2 = 1. \tag{10.34}$$

The flux of the outgoing wave, in the same units, is

$$|S\,e^{ikr}|^2 = |S|^2. \tag{10.35}$$

These two must be equal, in order that the state function should represent a continuous steady situation of beam and scatter—jet and splash—with no accumulation of probability density either at infinity or the origin. Therefore

$$|S|^2 = 1, \tag{10.36}$$

and S can be expressed in terms of a single real parameter δ, where

$$S = e^{2i\delta}. \tag{10.37}$$

Thus

$$u_S(r) \sim \frac{e^{2i\delta}\,e^{ikr} - e^{-ikr}}{2ikr}, \tag{10.38}$$

$$\equiv \frac{e^{ikr} - e^{-ikr}}{2ikr} + \left(\frac{e^{2i\delta} - 1}{2ik}\right)\frac{e^{ikr}}{r}. \tag{10.39}$$

Comparing (10.38) and (10.31), it is evident that the effect of the scattering potential is to shift the phase of the outgoing wave, relative to the incoming wave in the S-wave part of the original plane beam. For this reason the parameter δ is known as the *phase shift*.

Comparing (10.32) and (10.39),

$$f = \frac{e^{2i\delta} - 1}{2ik} = \frac{e^{i\delta}}{k}\left(\frac{e^{i\delta} - e^{-i\delta}}{2i}\right)$$

$$= (e^{i\delta}\sin\delta)/k. \tag{10.40}$$

Thus the S-wave scattering can be expressed in terms of the (real) phase shift δ. By (10.26),

$$\frac{d\sigma}{d\Omega} = \sigma(\theta) = \frac{\sin^2\delta}{k^2}, \tag{10.41}$$

which is an isotropic distribution. The total S-wave cross-section is

$$\sigma_S = \frac{4\pi \sin^2 \delta}{k^2},$$

$$= \pi \left(\frac{2 \sin \delta}{k}\right)^2. \tag{10.42}$$

Thus $2 \sin \delta/k$ is the effective radius of the target. Since

$$\sin \delta \leqslant 1,$$

$$\sigma_S \leqslant \frac{4\pi}{k^2}, \tag{10.43}$$

an upper limit for S-wave scattering at an energy corresponding to k, which is purely geometric and quite independent of the interaction potential.

Note that the above argument that the $l = 0$ scattering can be expressed in terms of a single phase shift is quite general, and does not depend on the detailed form of $V(r)$ (provided only that it falls off faster than $1/r$ for large r). For a given $V(r)$, δ is determined by the fact that (10.38) is the asymptotic form of the solution to the Schrödinger equation. For a given energy this determines a particular value of δ, in much the same way that the boundary conditions determine the energy levels of a bound system.

The entire scattering from a given potential, at a given energy, can be described in terms of a set of phase shifts, $\delta_l(k)$, one for each angular momentum state, l. However only those δ_l corresponding to l which satisfy (10.28) will be appreciably different from zero.

A particularly simple case for which δ can be solved exactly is that of the S-wave scattering from a hard sphere. The potential is given in (10.17). According to (7.10), if we put

$$\chi_S(r) = r u_S(r),$$

then for $r > a$, χ_S satisfies

$$\left[\frac{-\hbar^2}{2m}\frac{\partial^2}{\partial r^2} - E\right]\chi_S(r) = 0, \qquad E = \frac{\hbar^2 k^2}{2m},$$

or

$$\left[\frac{\partial^2}{\partial r^2} + k^2\right]\chi_S(r) = 0. \tag{10.44}$$

The solution must be of the form (10.38) (apart from the factor of r) and must also satisfy the hard-sphere boundary condition (see (3.23)), that

$$\chi_S(a) = 0. \tag{10.45}$$

Thus

$$\delta = -ka, \tag{10.46}$$

and the total S-wave scattering is

$$\sigma_S = \frac{4\pi}{k^2} \sin^2 ka. \tag{10.47}$$

If we are in the extreme low energy limit,

$$ka \ll 1,$$

then only S-wave scattering takes place (see (10.29)) and this is the total scattering. Thus

$$\sigma \simeq \frac{4\pi}{k^2}(ka)^2 = 4\pi a^2. \tag{10.48}$$

Note that this is larger, by a factor of four, than the classical cross section for point particles off a hard sphere of the same radius (Eq. (10.13)).

§ 10.5 Laboratory and Centre of Mass Systems

In the whole of the above discussion we have assumed that a beam of particles is being scattered by some fixed scattering centre. In fact the target is always free to move and the theory must be corrected to allow for this.

The technique for doing this is discussed in § 7.5 in connection with the hydrogen atom. The motion of two particles moving under their mutual interaction can always be expressed in terms of the free motion of the centre of gravity, and the relative motion of the two particles, which is controlled by their mutual interaction potential. It is the latter motion which is physically interesting. As was demonstrated in § 7.5 this is described by the equations considered above, provided only that we interpret the mass m as the reduced mass,

$$\frac{1}{m} \to \frac{1}{\mu} = \frac{1}{m_1} + \frac{1}{m_2} \tag{10.49}$$

(where m_1 is the mass of the beam particles, and m_2 that of the target), and take p to be the momentum of the beam in the co-ordinate frame in which the total momentum of the system is zero. Since in this co-ordinate system the centre of mass is at rest, it is called the Centre of Mass (C.M.) frame.

The experiment is actually carried out in the laboratory, where the beam particles have momentum q, say, and the target is at rest. This is called the laboratory (lab.) frame. In order to compare experiment with theory, it is necessary to convert what is actually observed in the lab. frame to what would have been observed in the C.M. frame. Since the observation of scattered flux is done with macroscopic apparatus, this is a purely classical problem and can be solved using classical notions.

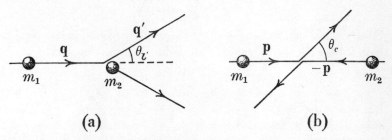

(a) (b)

FIG. 10.5. A typical collision showing the scattering angle and momenta in the (a) Laboratory and (b) Centre of Mass systems.

A typical collision as it would appear in the lab. and C.M. systems is illustrated in Fig. 10.5. In the lab. frame the total momentum of the system is q. Thus the velocity of the centre of mass is v, where

$$v = \frac{q}{m_1 + m_2}, \tag{10.50}$$

and the C.M. frame is moving in the beam direction, with velocity v, with respect to the lab. frame. The relation between the magnitudes of the momenta of the beam particles in the two frames is therefore

$$p = q - m_1 \left(\frac{q}{m_1 + m_2} \right) = \frac{m_2 q}{m_1 + m_2} \tag{10.51}$$

Since the total momentum is zero, this must also be the magnitude of the momentum of the target particle in the C.M. frame. (This follows immediately from (10.50). By energy conservation, it is also the magnitude of the momenta of both particles after the collision. If $\mathbf{q}'$ is the final momenta of the beam particle in the lab. frame, we have

9

from Fig. 10.5 (similarly to (10.51)) for the final momenta of the beam particle in the initial direction

$$q' \cos \theta_l = p \cos \theta_c + \frac{m_1 q}{m_1 + m_2} = p \cos \theta_c + \frac{m_1}{m_2} p. \qquad (10.52)$$

The final transverse momentum is the same in the two frames. Thus

$$q' \sin \theta_l = p \sin \theta_c. \qquad (10.53)$$

From the ratio of these two equations

$$\tan \theta_l = \frac{\sin \theta_c}{\cos \theta_c + (m_1/m_2)}. \qquad (10.54)$$

Clearly the azimuthal angle about the beam direction is the same in both frames

$$\phi_l = \phi_c. \qquad (10.55)$$

Suppose θ_l and ϕ_l specify a certain solid angle $d\Omega(\theta_l, \phi_l)$ in the lab. frame, and that $d\Omega(\theta_c, \phi_c)$ is the description of the same physical solid angle in the C.M. frame. If certain physical particles are observed to be scattered into $d\Omega(\theta_l, \phi_l)$ during a certain time interval in the lab. frame, the same particles would be seen to be scattered into $d\Omega(\theta_c, \phi_c)$ in the C.M. frame. Since the flux of the beam relative to the target is unchanged by a uniform velocity given to the whole system, we must have

$$\sigma(\theta_c, \phi_c) \, d\Omega(\theta_c, \phi_c) = \sigma(\theta_l, \phi_l) \, d\Omega(\theta_l, \phi_l), \qquad (10.56)$$

or,

$$\sigma(\theta_l, \phi_l) = \sigma(\theta_c, \phi_c) \frac{\sin \theta_c}{\sin \theta_l} \frac{d\theta_c}{d\theta_l} \frac{d\phi_c}{d\phi_l}. \qquad (10.57)$$

It then follows from (10.54) and (10.55) after some algebra that

$$\sigma(\theta_l, \phi_l) = \frac{[1 + (m_1/m_2)^2 + 2(m_1/m_2) \cos \theta_c]^{3/2}}{[1 + (m_1/m_2) \cos \theta_c]} \sigma(\theta_c, \phi_c). \qquad (10.58)$$

The kinetic energy of the system in the lab. frame is

$$T_l = \frac{q^2}{2m_1}. \qquad (10.59)$$

In the C.M. frame it is

$$T = \frac{p^2}{2m_1} + \frac{p^2}{2m_2} = \frac{p^2}{2\mu}. \qquad (10.60)$$

From (10.51) it follows that

$$T_c = \frac{m_2}{m_1 + m_2} T_l.$$ (10.61)

It is clear from all the above equations that the lab. frame is equivalent to the C.M. frame in the limit $m_2 \to \infty$, and that neglect of the recoil of the target is only a reasonable approximation if

$$\frac{m_1}{m_2} \ll 1.$$ (10.62)

The distinction between the two frames is, of course, essential when beam and particle masses are equal, as in the case of proton–neutron scattering. Then

$$T_c = \tfrac{1}{2} T_l.$$ (10.63)

§ 10.6 Summary

We have developed both the classical and quantum mechanical theories of the scattering of particles in a beam by a target particle, from which one may obtain information about their mutual interaction. The nature of the argument is illustrated in § 10.2 by comparing the effects of two extreme types of interaction potential.

For the quantum theory, the most important results are contained in Eqs. (10.40)–(10.43), which show how the S-wave ($l = 0$) scattering at any given energy, by any potential, may be expressed in terms of a single real parameter, δ, the *phase shift*.

PROBLEMS X

10.1. The impact parameter of a trajectory corresponding to a deflection θ is

$$b(\theta) = a \cos^2 (\theta/2).$$

Find the differential cross-section and the total cross-section.

10.2. The conservation of probability requires that S defined in (10.33) satisfies the relation

$$SS^* = 1.$$

By expressing f in terms of S show that this implies

$$f - f^* = 2ikff^*,$$

and, hence, that for S-wave scattering

$$4\pi \operatorname{Im} f = k\sigma_{\text{tot}}.$$

CHAPTER 11

THE NUCLEON–NUCLEON INTERACTION

§ 11.1 The Deuteron

Just as our deepest understanding of atomic structure is obtained from a very detailed study of the simplest atom, hydrogen, so the most direct information of the nucleon–nucleon interaction comes from the study of the two-nucleon system. The majority of the information comes from nucleon–nucleon scattering experiments, but fortunately there exists a bound state of the neutron–proton system—the deuteron—which is the nucleus of heavy hydrogen or deuterium. A consideration of this system gives very valuable information on the nature of the nucleon–nucleon potential.

We have already seen that the size of a large nucleus is about 10^{-12} cm, and that the nuclear potential is a short-range affair interacting over not more than this distance. If the nuclear potential is of sufficiently long range for each pair of the A nucleons in a nucleus to interact with each other, the binding energy should be expected to vary with $[A(A-1)]/2$, since this is the number of interacting pairs. In fact, for heavy nuclei, for which surface effects are small, the binding energy (and volume) vary with A. This suggests that the nuclear potential is of such short range that each nucleon only interacts with its nearest neighbours, and we estimate the range, on the basis of (9.36), to be

$$a \simeq 1 \cdot 5 \times 10^{-15} \text{ m}.$$

For simplicity we assume that the potential is central—depending only on the distance between the nucleons—and that, like the ground state of hydrogen, the deuteron has zero angular momentum. Equivalently, we may say that it is in an S-state.

For the hydrogen atom the potential is known, and the problem is to find the possible energy levels. In the case of the deuteron the binding energy is taken from experiment and used to put conditions on the shape of the potential. This binding energy, ϵ, is the energy required to separate the two nucleons, and is thus equal to the difference between the rest energy of the proton and neutron, and that of the deuteron. In an obvious notation

$$\epsilon = (m_p + m_n)\,c^2 - m_d\,c^2. \tag{11.1}$$

Experimentally it is found that

$$\epsilon \simeq 2\ \text{MeV}. \tag{11.2}$$

If the proton and neutron in the deuteron are bound together by a potential of range a their distance apart must be of this order of

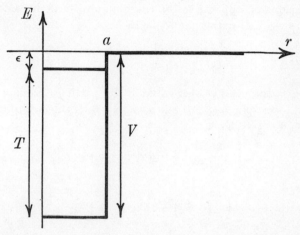

FIG. 11.1. The energy diagram of the deuteron in the "square well" approximation. A particle with binding energy ϵ, must have kinetic energy T, as shown, where $T + \epsilon = V$. If $T \gg \epsilon$ as a consequence of the uncertainty principle, it follows that $V \gg \epsilon$.

magnitude. By the uncertainty principle, this means that their relative momentum must be at least of order p, where (see Appendix)

$$p \simeq \hbar/a = \frac{1 \cdot 05 \times 10^{-34}}{1 \cdot 5 \times 10^{-15}}\ \text{mks}$$

$$\simeq 130\ \text{MeV/c}. \tag{11.3}$$

The mutual kinetic energy is

$$T = \frac{p^2}{2\mu}, \tag{11.4}$$

where μ is the reduced mass,

$$\frac{1}{\mu} = \frac{1}{m_p} + \frac{1}{m_n}\cdot \tag{11.5}$$

Therefore

$$T \simeq 20 \text{ MeV}. \tag{11.6}$$

The binding energy, ϵ, is the difference between the (negative) potential energy, V, of the attractive potential which binds the particle together, and the kinetic energy, T, due to their relative motion. But ϵ is much smaller than T. From this it is evident (see Fig. 11.1), that

$$V \simeq T \gg \epsilon. \tag{11.7}$$

As a crude first approximation let us suppose that the potential is a "square well" of radius a, and depth V.

$$V(r) = -V \qquad r \leqslant a,$$
$$V(r) = 0 \qquad r > a. \tag{11.8}$$

Then $-\epsilon$ is the energy eigenvalue of the particles moving in this potential, corresponding to a state with $l = 0$. If the eigenfunction is $\chi(r)$ then, (see (7.10)),

$$\left[\frac{-\hbar^2}{2\mu} \frac{\partial^2}{\partial r^2} + V(r) \right] \chi(r) = -\epsilon \chi(r). \tag{11.9}$$

Thus

$$\left(\frac{\partial^2}{\partial r^2} - K^2 \right) \chi = 0, \qquad r > a. \tag{11.10}$$

where

$$K^2 = \frac{2\mu\epsilon}{\hbar^2}, \tag{11.11}$$

and

$$\left(\frac{\partial^2}{\partial r^2} + k^2 \right) \chi = 0, \qquad r \leqslant a \tag{11.12}$$

where

$$k^2 = \frac{2\mu(V-\epsilon)}{\hbar^2} \simeq \frac{2\mu V}{\hbar^2}. \tag{11.13}$$

χ must tend to zero at infinity and χ/r must be finite at the origin; hence the solution is

$$\chi = A \sin kr, \qquad r \leqslant a, \tag{11.14}$$

$$\chi = B e^{-Kr}, \qquad r > a. \tag{11.15}$$

Since χ and χ' are continuous at $r = a$,

$$A \sin ka = B e^{-Ka}, \tag{11.16}$$

$$kA \cos ka = -KB e^{-Ka}. \tag{11.17}$$

Taking the ratio of these two equations (compare with (4.50)),

$$k \cot ka = -K, \tag{11.18}$$

which is the eigenvalue equation, which would determine the value of ϵ, if V and a were known. Using the approximation (11.13),

$$\cot ka = -\frac{K}{k} \simeq -\left(\frac{\epsilon}{V}\right)^{1/2}, \tag{11.19}$$

which is small. Thus

$$ka \simeq \pi/2. \tag{11.20}$$

From (11.20) and the definition (11.13),

$$k^2 = \frac{\pi^2}{4a^2} = \frac{2\mu V}{\hbar^2}, \tag{11.21}$$

or

$$a^2 V = \frac{\pi^2 \hbar^2}{4m_n}. \tag{11.22}$$

This relation between the depth and radius is the condition imposed on the potential by the experimental fact that the deuteron binding energy is small. Its precise form depends on the shape we have chosen for the potential, but if any other shape had been chosen a similar condition on the "volume", $a^2 V$, of the potential would have been obtained.

The relation can be considered from a different point of view. The state function of the deuteron in the region $r > a$ falls off like $\exp[-rK]$. We may thus define the radius of the deuteron as $1/K$, since this is a measure of the range of values of r for which there is an appreciable probability of finding the particles. From the definition, (11.11),

$$1/K = \frac{\hbar}{(m_n \epsilon)^{1/2}}. \tag{11.23}$$

But by the weak binding condition, (11.22), the radius of the potential is

$$a = \frac{\pi}{2} \frac{\hbar}{(m_n V)^{1/2}}. \tag{11.24}$$

Thus

$$\frac{(1/K)}{a} = \frac{2}{\pi}\left(\frac{V}{\epsilon}\right)^{1/2}, \tag{11.25}$$

which is a number greater than one. From (11.2), (11.6) and (11.7), it is probably about three. This shows that the radius of the deuteron is considerably greater than the radius of the potential. It is a very loosely bound structure in which the particles spend a large fraction of their time outside the range of the attractive potential which binds them together.

§ 11.2 Neutron–Proton Scattering

We now consider the scattering of a beam of neutrons by protons in the low energy region, in which only scattering in the $l = 0$ state takes place. We show that this can be expressed in terms of the binding energy of the deuteron.

Define

$$U(r) = \frac{2\mu}{\hbar^2} V(r), \tag{11.26}$$

where $V(r)$ is assumed to vanish outside some finite radius. Then, if the energy of the particles in the centre of mass frame is $\hbar^2 k_1^2/2\mu$, the state function χ_1 satisfies the equation, (7.10),

$$\left[\frac{\partial^2}{\partial r^2} + k_1^2 - U(r)\right]\chi_1(r) = 0. \tag{11.27}$$

Similarly, for a different momentum,

$$\left[\frac{\partial^2}{\partial r^2} + k_2^2 - U(r)\right]\chi_2(r) = 0. \tag{11.28}$$

Multiply (11.27) by χ_2 and (11.28) by χ_1 and subtract. The first two terms are

$$\chi_2 \frac{\partial^2}{\partial r^2}\chi_1 - \chi_1 \frac{\partial^2}{\partial r^2}\chi_2 \equiv \frac{\partial}{\partial r}\left(\chi_2 \frac{\partial \chi_1}{\partial r} - \chi_1 \frac{\partial \chi_2}{\partial r}\right). \tag{11.29}$$

Thus

$$\frac{\partial}{\partial r}\left(\chi_2 \frac{\partial \chi_1}{\partial r} - \chi_1 \frac{\partial \chi_2}{\partial r}\right) + \chi_1 \chi_2 (k_1^2 - k_2^2) = 0. \tag{11.30}$$

If this equation is integrated between 0 and some large radius, R, then

$$\left[\chi_2 \frac{\partial \chi_1}{\partial r} - \chi_1 \frac{\partial \chi_2}{\partial r}\right]_0^R = (k_2^2 - k_1^2) \int_0^R \chi_1 \chi_2 \, dr. \tag{11.31}$$

If the asymptotic form of $\chi(r)$ for large r is $\phi(r)$, then by precisely the same argument,

$$\left[\phi_2 \frac{\partial \phi_1}{\partial r} - \phi_1 \frac{\partial \phi_2}{\partial r}\right]_0^R = (k_2^2 - k_1^2) \int_0^R \phi_1 \phi_2 \, dr, \tag{11.32}$$

since ϕ satisfies the same equation as χ, except that $U(r)$ is replaced by zero. Subtract (11.32) from (11.31), and evaluate for $R \to \infty$. For large R, $\phi(R) = \chi(R)$, and $\chi(0)$ is zero. Thus

$$\left[\phi_2 \frac{\partial \phi_1}{\partial r} - \phi_1 \frac{\partial \phi_2}{\partial r}\right]_{r=0} = (k_2^2 - k_1^2) \int_0^\infty (\chi_1 \chi_2 - \phi_1 \phi_2)\, dr. \quad (11.33)$$

If k_1^2 is positive, the state represents scattering and, according to (10.38), is

$$\phi_{k_1}^{(r)} = \phi_k^{(r)} \sim \left[\frac{e^{i(kr+\delta)} - e^{-i(kr+\delta)}}{2ik}\right] e^{i\delta}, \quad (11.34)$$

$$= \frac{\sin (kr+\delta)}{\sin \delta}, \quad (11.35)$$

where, for convenience, we have chosen to normalize so that

$$\phi_k(0) = 1. \quad (11.36)$$

If k_2^2 is negative, we have a bound state. We can put

$$k_2^2 = -K^2, \quad (11.37)$$

which is the notation of the previous section. Thus

$$\phi_{k_2} = \phi_K = e^{-Kr}, \quad (11.38)$$

which is also normalized so that

$$\phi_K(0) = 1. \quad (11.39)$$

Since the potential is of short range, the asymptotic forms, ϕ, are equal to the corresponding exact state functions over most of the range of integration in the integral in (11.33), and to first approximation this term can be neglected. Substituting (11.35) and (11.38) into (11.33), in this approximation, gives

$$k \cot \delta = -K. \quad (11.40)$$

Since K is directly related to the binding energy, (11.11), this equation determines the S-wave scattering phase shift, and hence the low-energy scattering, in terms of the deuteron binding. For small k,

$$\delta \simeq -k/K, \quad (11.41)$$

which shows by comparison with (10.46) that the deuteron "radius", $1/K$, is also the radius of the equivalent hard sphere for the extreme low-energy scattering. In this same approximation

$$\sigma = \frac{4\pi \sin^2 \delta}{k^2}$$

$$= \frac{4\pi \delta^2}{k^2}$$

$$= \frac{4\pi}{K^2}$$

$$= \frac{4\pi \hbar^2}{m_n \epsilon}. \tag{11.42}$$

Substituting the experimental values, this gives a total cross-section of about 2×10^{-24} cm^2 or 2 barns. The experimental cross-section is about 50 barns. Since the argument leading to (11.42) is extremely general, depending only on the assumption of a short-range potential, this discrepancy is at first sight rather startling. The explanation is quite simple. The nucleons like the electrons, have spin and if the effect of this is included everything can be explained quite simply.

§ 11.3 Spin-dependent Interaction

If the proton and neutron each have spin $\frac{1}{2}$ (in units of $\hbar$), classically the total spin of the proton–neutron system may be anything from zero to one, depending on the relative orientation of the two spin vectors. The maximum and minimum values come from the parallel and anti-parallel configurations, respectively. Quantum mechanically only the integer values 0 and 1 are allowed.

To confirm this it is sufficient to count the number of independent spin states, which must be the same, whether the states are specified by the total spin and its orientation, or by the individual spin orientations. The individual proton and neutron spins may be "up" or "down" with respect to some arbitrarily chosen direction, giving rise to the states called $|+\frac{1}{2}\rangle$ and $|-\frac{1}{2}\rangle$ in § 8.3. There are now four independent states which we denote by

$$|+\tfrac{1}{2}\rangle_p |+\tfrac{1}{2}\rangle_n, \quad |+\tfrac{1}{2}\rangle_p |-\tfrac{1}{2}\rangle_n, \quad |-\tfrac{1}{2}\rangle_p |+\tfrac{1}{2}\rangle_n, \quad |-\tfrac{1}{2}\rangle_p |-\tfrac{1}{2}\rangle_n$$

in an obvious notation. If the total spin is j there are $(2j+1)$ independent states (see Chapter 6). Thus if there are three states of total spin 1, and one state of total spin zero, this again gives the correct total of four.

If we denote the state of total spin j and z-component m by $|j,m\rangle$, the two ways of specifying the spin states are related as follows.

$$|1, +1\rangle = |+\tfrac{1}{2}\rangle_p |+\tfrac{1}{2}\rangle_n$$

$$|1, \ \ \ 0\rangle = (\tfrac{1}{2})^{1/2}[\,|+\tfrac{1}{2}\rangle_p|-\tfrac{1}{2}\rangle_n + |-\tfrac{1}{2}\rangle_p|+\tfrac{1}{2}\rangle_n], \qquad (11.43)$$

$$|1, -1\rangle = |-\tfrac{1}{2}\rangle_p|-\tfrac{1}{2}\rangle_n,$$

$$|0, \ \ \ 0\rangle = (\tfrac{1}{2})^{1/2}[\,|+\tfrac{1}{2}\rangle_p|-\tfrac{1}{2}\rangle_n - |-\tfrac{1}{2}\rangle_p|+\tfrac{1}{2}\rangle_n].$$

The $|1, +1\rangle$ and $|1, -1\rangle$ states are obvious, since it is only by having the individual spins aligned that a total z-component of ± 1 can be obtained. The $|1,0\rangle$ configuration then follows, since it must share with $|1, +1\rangle$ and $|1, -1\rangle$ the property of being symmetric under exchange of $p \leftrightarrow n$. The $|0,0\rangle$ state must then be the antisymmetric combination, since this is orthogonal to all the others.

Since the neutron and proton have spin, we must introduce Pauli spin operators, $\hat{\boldsymbol{\sigma}}_p$ and $\hat{\boldsymbol{\sigma}}_n$†. The interaction potential between the two particles may depend on these spin operators, just as the potential energy of an electron in a magnetic field depends on $\hat{\boldsymbol{\sigma}}.\boldsymbol{B}$ (see (8.52)). The potential energy must be a scalar, so the spin operators, which form an axial vector must appear as a scalar product with some other axial-vector. An obvious possibility is $(\hat{\boldsymbol{\sigma}}_n + \hat{\boldsymbol{\sigma}}_p).\hat{\boldsymbol{l}}$, where the components of $\hat{\boldsymbol{l}}$ are the orbital angular momentum operators (6.2). However, for a state for which $l = 0$, this will make no contribution. The simplest term, which is non-vanishing for our problem, is an interaction potential proportional to $\hat{\boldsymbol{\sigma}}_p.\hat{\boldsymbol{\sigma}}_n$.

To calculate the effect of this operator on states of total spin $j = 1, |1\rangle$, and $j = 0, |0\rangle$, we have the relation between the operators

$$\hat{\boldsymbol{J}} = \tfrac{1}{2}(\hat{\boldsymbol{\sigma}}_p + \hat{\boldsymbol{\sigma}}_n). \qquad (11.44)$$

Thus

$$\hat{\boldsymbol{\sigma}}_p.\hat{\boldsymbol{\sigma}}_n = 2[\hat{\boldsymbol{J}}^2 - (\tfrac{1}{2}\hat{\boldsymbol{\sigma}}_p)^2 - (\tfrac{1}{2}\hat{\boldsymbol{\sigma}}_n)^2]. \qquad (11.45)$$

Hence, operating on the eigenstate $|j\rangle$,

$$\hat{\boldsymbol{\sigma}}_p.\hat{\boldsymbol{\sigma}}_n|j\rangle = 2\{j(j+1) - \tfrac{1}{2}(\tfrac{1}{2}+1) - \tfrac{1}{2}(\tfrac{1}{2}+1)\}|j\rangle. \qquad (11.46)$$

This follows from (11.43), since the operator $\hat{\boldsymbol{\sigma}}_p$ operates only on the eigenstates $|\pm\tfrac{1}{2}\rangle_p$ and $\hat{\boldsymbol{\sigma}}_n$ on $|\pm\tfrac{1}{2}\rangle_n$. Hence, substituting for j,

$$\hat{\boldsymbol{\sigma}}_p.\hat{\boldsymbol{\sigma}}_n|1\rangle = |1\rangle, \qquad (11.47)$$

and

$$\hat{\boldsymbol{\sigma}}_p.\hat{\boldsymbol{\sigma}}_n|0\rangle = -3|0\rangle. \qquad (11.48)$$

† The vector operator $\hat{\boldsymbol{\sigma}}$ has components $(\hat{\sigma}_x, \hat{\sigma}_y, \hat{\sigma}_z)$.

We now assume that the interaction potential between the neutron and proton is

$$= V_C(r) + \hat{\boldsymbol{\sigma}}_n . \hat{\boldsymbol{\sigma}}_p \, V_S(r). \tag{11.49}$$

The state function, in addition to the space dependent part, must also include a factor specifying the spins of the nucleons. It is convenient to specify this in terms of the total spin, by one of the spin state vectors, $|j\rangle$. For scattering in an $l = 0$ state the appropriate Schrödinger equation is

$$\left[-\frac{\hbar^2}{2\mu} \frac{\partial^2}{\partial r^2} + V_C(r) + \hat{\boldsymbol{\sigma}}_n . \hat{\boldsymbol{\sigma}}_p \, V_S(r) - \frac{\hbar^2 k^2}{2\mu} \right] \chi_j(r) |j\rangle = 0. \tag{11.50}$$

For $|j\rangle = |1\rangle$, we have, by (11.47),

$$\left[-\frac{\hbar^2}{2\mu} \frac{\partial^2}{\partial r^2} + V_C(r) + V_S(r) - \frac{\hbar^2 k^2}{2\mu} \right] \chi_1(r) = 0, \tag{11.51}$$

where we have cancelled the spin vector $|1\rangle$ from the state function, since there is no longer any spin dependence in the operator. Similarly, by (11.48), for $|j\rangle = |0\rangle$,

$$\left[-\frac{\hbar^2}{2\mu} \frac{\partial^2}{\partial r^2} + V_C(r) - 3V_S(r) - \frac{\hbar^2 k^2}{2\mu} \right] \chi_0(r) = 0. \tag{11.52}$$

The deuteron has spin 1 and is thus a bound state solution of (11.51). By the argument given in the previous section it determines the low-energy scattering, σ_1, in the effective potential for spin one states,

$$V_1(r) = V_C(r) + V_S(r). \tag{11.53}$$

However, the low-energy scattering for spin zero states, σ_0, is determined by the completely independent effective potential of (11.52), namely

$$V_0(r) = V_C(r) - 3V_S(r). \tag{11.54}$$

Since there are three states of spin one and one state of spin zero, all equally probable, the observed low-energy proton–neutron cross-section is

$$\sigma = \tfrac{3}{4}\sigma_1 + \tfrac{1}{4}\sigma_0. \tag{11.55}$$

The deuteron binding energy determines σ_1, but gives no information on σ_0, and the apparent paradox of the previous section is resolved.

The question of the nucleon potential will not be considered further in detail. Experiments with beams of protons of higher and higher energy have led on to much more interesting problems, which are outlined in the next section.

§ 11.4 Outline of Further Developments

The problem of determining the nucleon–nucleon interaction potential has still not been solved. Instead, experiments in this connection have uncovered a whole new field of research.

We have discussed above the theory of proton–neutron scattering. This is a difficult experiment to perform directly, since one cannot have targets consisting only of neutrons. Proton–proton collisions are far easier to set up, since protons can be accelerated artificially, and the beams fired into targets of liquid hydrogen. If the beam energies are low the protons are pushed apart by the Coulomb repulsion before they get close enough for the short-range nuclear potential to operate, as in the Rutherford experiment. The collisions are then controlled by the well-known electrical effects and no direct information is obtained about the nuclear interaction. Thus the first experimental requirement is for proton beams of higher and higher energies, which can penetrate deeper and deeper into the region of specifically nuclear interaction.

As the energy is increased nothing startling happens until it reaches about 300 MeV in the laboratory (equivalent, according to (10.63), to 150 MeV in the C.M. system), when a qualitative change takes place. The protons, instead of emerging from the collision like a couple of billiard balls with conservation of kinetic energy, may come off quite slowly accompanied by a third particle—a π-meson.

In the first place this is a direct and dramatic example of the energy conservation law as modified by the theory of relativity. In the energy balance for the collision, the rest energy of the particles involved must be included, and kinetic energy and rest energy are convertable. The π-meson has mass about 140 MeV/c^2, and hence a rest energy of 140 MeV. As soon as this much kinetic energy is available in the C.M. system, π-meson creation in the p–p collision is energetically possible. The kinetic energy of the two initial protons is converted for the most part into the rest energy of the π-meson.

Although this reaction was first seen in 1947, it was not unexpected and had, in fact, been predicted as early as 1935. The electron and proton in the hydrogen atom interact through the Coulomb potential. When the theory of radiation is expressed in quantum mechanical terms, the radiation appears as photons. The Coulomb interaction may then be described as an exchange of photons between the two charged particles. They interact like two ball players, who throw a ball from one to the other. This sort of interaction arises naturally from the combination of quantum mechanics and relativity, and it was suggested by Yukawa† that the nuclear interaction might be of the same type. The range of such an interaction is related to the mass of the exchanged particle. (The ball players can throw a tennis ball over a large distance, but have to be very close before they can swap cannon balls.) The range of interaction associated with the exchange of a particle of mass m in a relativistic quantum theory, must, on dimensional grounds, be the length which can be constructed from m and the other relevant constants, $\hbar$ and c. This is

$$R = \frac{\hbar}{mc} \cdot \qquad (11.56)$$

If we take

$$R \simeq 1 \cdot 5 \times 10^{-15} \text{ m}, \qquad (11.57)$$

in accordance with (9.36), the rest energy of the related particle is

$$mc^2 \simeq 130 \text{ MeV}. \qquad (11.58)$$

A consequence of Yukawa's idea is that these particles should be produced in nucleon–nucleon collisions if sufficient energy is available. The appearance of π-mesons with rest energy close to this value was thus direct confirmation of the Yukawa theory.

However, if the energy of the proton–proton collision is increased further, many more sub-nuclear—or elementary—particles are produced. These particles were totally unexpected. At even higher energies the corresponding anti-particles start to appear. These bear the same relation to the particles as positrons do to electrons. When anti-particles subsequently collide with particles they "annihilate", converting their mass rapidly back into lighter particles and kinetic energy.

The total number of particles and anti-particles now known is over one hundred. Those well established by about 1960 are shown in Table 11.1. It will be observed that most of them are unstable, and decay into other particles with mean lives varying from 10^{-6} to 10^{-10} sec. Some of them such as muons, μ, and neutrinos, ν, are not produced directly in nucleon–nucleon collisions, but only appear as decay products of the others. The lifetimes of the unstable particles are very long on the nuclear time scale

$$(\hbar/m_p c^2 \simeq 10^{-24} \text{ sec}), \qquad (11.59)$$

so the interactions which cause their disintegrations are extremely weak (see § 14.4). There are thus two types of short-range nuclear

† H. Yukawa, *Proc. Phys. Math. Soc. Japan*, **17**, 48 (1935).

interaction—the *"strong"* *interactions*, which bind the nucleons together, and produce π-mesons and other particles in nucleon–nucleon collisions; and the *"weak"* *interactions*, which cause these particles to disintegrate—or decay. These two types of nuclear interaction have to be added to the two long-range interactions known to classical physics—the electromagnetic interactions of Maxwell, and the gravitational interaction of Newton.

TABLE 11.1 *The elementary particles*

The upper suffices denote electric charge. The particles appear in pairs of opposite charge, known as particle and anti-particle. Masses are given to the nearest 5 MeV/c², and mean lives to an order of magnitude in seconds. The table summarizes the situation as it was known in about 1960. More recent developments are discussed in Part IV.

	Particle	Decay products	Mean life (sec)	Mass (MeV/c²)	Spin $\hbar$	Anti-particle
Baryons	Ξ^-	$\rightarrow \Lambda + \pi^-$	10^{-10}	1320	$\frac{1}{2}$	$\overline{\Xi}^+$
	Ξ^0	$\rightarrow \Lambda + \pi^0$	10^{-10}	1315	$\frac{1}{2}$	$\overline{\Xi}^0$
	$\Sigma^\pm$	$\rightarrow n + \pi^+, p^+ + \pi^\circ$	10^{-10}	1190	$\frac{1}{2}$	$\overline{\Sigma}^\mp$
	Σ^0	$\rightarrow \Lambda + \gamma$	10^{-18}	1190	$\frac{1}{2}$	$\overline{\Sigma}^0$
	Λ^0	$\rightarrow p^+ + \pi^-$	10^{-10}	1115	$\frac{1}{2}$	$\overline{\Lambda}^0$
	n	$\rightarrow p + e^- + \bar{\nu}$	10^3	939	$\frac{1}{2}$	$\bar{n}$
	p^+		stable	938	$\frac{1}{2}$	$\overline{p}^-$
Mesons	κ^+	$\rightarrow 2\pi, 3\pi; \mu + \nu$	10^{-8}	495	0	κ^-
	κ^0	$\rightarrow 2\pi, 3\pi$	10^{-10}	500	0	κ^0
	η^0	$\rightarrow 2\gamma, 3\pi$	10^{-16}	560	0	η^0
	π^+	$\rightarrow \mu^+ + \nu$	10^{-8}	140	0	π^-
	π^0	$\rightarrow 2\gamma$	10^{-16}	135	0	π^0
Photon	γ		stable	0	1	γ
Leptons (muon)	μ^-	$\rightarrow e^- + \nu + \nu$	10^{-6}	105	$\frac{1}{2}$	μ^+
(electron)	e^-		stable	$\frac{1}{2}$	$\frac{1}{2}$	e^+
(neutrino)	ν_e		stable	0	$\frac{1}{2}$	$\bar{\nu}_e$
(neutrino)	ν_μ		stable	0	$\frac{1}{2}$	$\bar{\nu}_\mu$

If dimensionless constants are introduced, analogous to the fine structure constant, (8.60), to define the strengths of the four fundamental interactions, they are as shown in Table 11.2.

The two Tables, (11.1) and (11.2), describe the raw material of physics as we now know it. The whole of the universe, both the matter and the radiation, is made up of the thirty-odd particles listed in Table 11.1. Everything that happens, which can be described in physical or chemical terms, is a consequence of one of the four types of interaction listed in Table 11.2. It is a remarkable synthesis, but is far more complicated than one might expect, and there is a great

TABLE 11.2. *The fundamental interactions of nature, showing strength and range*

Interaction	Strength	Range of potential
Electro-magnetic	$e_M^2/\hbar c = \frac{1}{137}$	∞, ($1/r$ law)
Gravitational	$\gamma m_p^2/\hbar c \simeq 10^{-36}$	∞, ($1/r$ law)
Strong (nuclear)	$g^2/\hbar c \simeq 1$	short ($\sim 10^{-15}$ m)
Weak (nuclear)	$(f^2/\hbar c)^2 \simeq 10^{-10}$	short ($\sim 10^{-15}$ m)

deal which we do not begin to understand. The nucleon–nucleon potential now appears as just one aspect of the "strong" nuclear interaction. The complicated pattern of the particle decay schemes, listed in Table 11.1, all arises from the "weak" interaction. We have no detailed knowledge of either "strong" or "weak" interactions, but very recently considerable progress has been made in classifying the strongly interacting particles. This is described in Chapter 15.

PART IV

General Theory
and
Sub-nuclear Physics

CHAPTER 12

OPERATORS AND STATE VECTORS

§ 12.1 Dirac Notation

In Chapter 3 we show the physical necessity for introducing operators to represent the operations of measurement on quantum systems, and develop a scheme for relating the mathematics of operators to the observations of quantum physics. In order not to obscure the main argument, we there gloss over a number of essentially mathematical points. We return to these now. In particular we show how the physical interpretation of the overlap integral, which is introduced as as extra postulate in (3.39), can be derived from the other interpretive assumptions of § 3.2.

To understand the new mathematical structure of quantum theory it is an enormous assistance to introduce a notation due to Dirac. In Chapter 8 it was found necessary, in order to describe electron spin, to generalize the theory to include the possibility of matrix operators and state vectors, playing an analogous role to differential operators and state functions. The Dirac notation is designed, *inter alia*, to exploit this analogy to the full. We have already introduced this notion for the spin vectors in Chapters 8 and 11, and we now cast the theory of state functions into a similar form.

In all the above work a general state function, not an eigenfunction of any observable, has been written $\psi(x)$. In the Dirac notation we write this as

$$\psi(x) \equiv \langle x|\psi \rangle. \tag{12.1}$$

We also introduce the notation for the complex conjugate function,

$$\psi^*(x) \equiv \langle \psi|x \rangle. \tag{12.2}$$

It is often convenient not to mention the x dependence explicitly and refer simply to $|\psi\rangle$. In this form it is more exact to call $|\psi\rangle$ a state vector, rather than a state function. It plays a role in configuration space exactly analogous to the spin vectors $|\psi\rangle$ introduced in § 8.2. The spin vector has two components designated by $\langle 1|\psi\rangle$, $\langle 2|\psi\rangle$ (see (8.13)), or in the more sophisticated version (see (8.45)) $\langle +\frac{1}{2}|\psi\rangle$ and $\langle -\frac{1}{2}|\psi\rangle$. The expression $\langle x|\psi\rangle$ may be thought of, in the same way, as the x-component of a *state-vector* $|\psi\rangle$, only now we have an infinity of components, specified by the continuous variable x, which run together to form the *state function*, $\psi(x) \equiv \langle x|\psi\rangle$.

The normalizing condition, (3.32), can be written

$$\int \langle\psi|x\rangle\langle x|\psi\rangle \, dx = \int |\langle\psi|x\rangle|^2 dx = 1. \tag{12.3}$$

where, as usual, the integration is over the full physical range of the integration variable. In terms of state-vectors this can be abbreviated to

$$\langle\psi|\psi\rangle = 1. \tag{12.4}$$

If we have another general state function

$$\phi(x) = \langle x|\phi\rangle, \tag{12.5}$$

the overlap integral between the state vectors $|\phi\rangle$ and $|\psi\rangle$ (which occurs in (3.39) for $\phi = u_{a_n}$) is

$$\int \langle\phi|x\rangle\langle x|\psi\rangle \, dx \equiv \langle\phi|\psi\rangle. \tag{12.6}$$

The final expression is again a short hand for the integral. The Dirac notation shows clearly that the overlap integral is just the generalization of the scalar product of two (complex) vectors, (8.26), to the infinite continuous case. In vector language, if the overlap integral is zero,

$$\langle\phi|\psi\rangle = 0, \tag{12.7}$$

the two state-vectors are said to be orthogonal. Notice that it follows from the definitions (12.1), (12.2), and (12.6) that

$$\langle\phi|\psi\rangle = \langle\psi|\phi\rangle^*. \tag{12.8}$$

In the previous work, we have consistently denoted the eigenfunction of an operator $\hat{A}$, belonging to the eigenvalue a_n by $u_{a_n}(x)$. We now write this as

$$u_{a_n}(x) \equiv \langle x|a_n\rangle. \tag{12.9}$$

If it is unnecessary to refer explicitly to the x-dependence, we write the corresponding eigenvector as

$$u_{a_n} = |a_n\rangle. \tag{12.10}$$

This, again, is a generalization of the notation already introduced for finite spin vectors in (8.44). We could have written $|u_{a_n}\rangle$, but the form of bracket already tells us that we are dealing with a state-vector, and the eigenvalue label, a_n, tells us it is an eigenvector. The symbol u carries no additional information. It is convenient to drop it, and work with the abbreviated forms (12.9) and (12.10).

The eigenvalue equation, (2.7), reads

$$\hat{A}\left(x, \frac{\partial}{\partial x}\right)\langle x|a_n\rangle = a_n\langle x|a_n\rangle. \tag{12.11}$$

If we do not wish to refer explicitly to the x-dependence, this can be abbreviated to

$$\hat{A}|a_n\rangle = a_n|a_n\rangle, \tag{12.12}$$

which is a generalization of (8.16) for spin.

None of the actual calculations of eigenvalues and eigenfunctions presented in the previous chapters are simplified in any way by this change of notation, but it may be helpful to re-state a few results in the new language. Thus the eigenvalue equation for the z-component of angular momentum, (6.5), is

$$\hat{l}_z(\phi)\langle\phi|l_z\rangle = l_z\langle\phi|l_z\rangle, \tag{12.13}$$

or more formally

$$\hat{l}_z|l_z\rangle = l_z|l_z\rangle. \tag{12.14}$$

The normalized eigenfunctions are, (6.6),

$$\langle\phi|l_z\rangle = \frac{1}{\sqrt{2\pi}}\, e^{il_z\phi/\hbar}; \qquad l_z = \hbar[0, \pm1, \pm2, \ldots]. \tag{12.15}$$

The similar equations for total angular momentum are

$$\hat{l}^2(\theta,\phi)\,\langle\theta,\phi|l^2,l_z\rangle = \hbar^2 l(l+1)\langle\theta,\phi|l^2,l_z\rangle \tag{12.16}$$

where, (6.36),

$$l = 0, 1, 2, \ldots \tag{12.17}$$

and, (6.38),

$$\langle\theta,\phi|l^2,l_z\rangle = \langle\theta,\phi|l,m\rangle = Y_l^m(\theta,\phi). \tag{12.18}$$

The two labels l^2, l_z in $\langle\theta,\phi|l^2,l_z\rangle$ indicate that the eigenstates of $\hat{l}^2$ are degenerate; that they can be chosen to be simultaneous eigenstates of $\hat{l}^2$ and $\hat{l}_z$, and that the eigenvalues of both operators are required to specify a unique state function. This is further abbreviated, in an obvious notation, to the form $\langle\theta,\phi|l,m\rangle$. In the same way, the discrete eigenfunctions of the hydrogen atom, (7.34), can be written

$$u_{nlm}(r, \theta, \phi) = \langle r, \theta, \phi|E_n, l^2, l_z\rangle$$
$$\equiv \langle r, \theta, \phi|n, l, m\rangle. \tag{12.19}$$

The developments of Chapter 3 up to (3.7) can be written quite simply in the new notation, the advantage being that in terms of state vectors the results are equally applicable to differential or matrix operators. In particular (3.1), for the average value of repeated observations, $\hat{A}$, on a normalized state $|\psi\rangle$, reads

$$\bar{a}_\psi = \int \langle\psi|x\rangle \, \hat{A}\left(x, \frac{\partial}{\partial x}\right) \langle x|\psi\rangle \, dx$$

$$\equiv \langle\psi|\hat{A}|\psi\rangle. \tag{12.20}$$

§ 12.2 Observable Operators: Orthonormality

So far we have done nothing but change the notation. We now for the first time come to some refinements of the theory. For consistency, it is clearly necessary that (12.20) should be a real number. This implies some restriction on the operators $\hat{A}$, which can represent observations, which we must have assumed tacitly above. A necessary condition is clearly that

$$\langle\psi|\hat{A}|\psi\rangle = \langle\psi|\hat{A}|\psi\rangle^* \tag{12.21}$$

for any state $|\psi\rangle$. For reasons which appear below, we impose the more restrictive condition that:

an operator $\hat{A}$ cannot represent a physical observation unless

$$\langle\phi|\hat{A}|\psi\rangle = \langle\psi|\hat{A}|\phi\rangle^* \tag{12.22}$$

for any states $|\phi\rangle$ and $|\psi\rangle$.

In long-hand this restriction reads

$$\int \langle\phi|x\rangle \, \hat{A}\left(x, \frac{\partial}{\partial x}\right) \langle x|\psi\rangle \, dx = \left[\int \langle\psi|x\rangle \, \hat{A}\left(x, \frac{\partial}{\partial x}\right) \langle x|\phi\rangle \, dx\right]^*$$

Such an operator is called an *observable* operator. Clearly (12.22) has the desired effect of making (12.20) real. It has two other mathematical consequences embodied in the two following theorems, which have important physical implications.

THEOREM I. *The eigenvalues of an observable operator are real.*
Proof. If a is any eigenvalue of $\hat{A}$ then

$$\hat{A}|a\rangle = a|a\rangle, \tag{12.23}$$

$$\hat{A}\left(x, \frac{\partial}{\partial x}\right) \langle x|a\rangle = a\langle x|a\rangle.$$

Therefore, multiplying by $\langle a|x\rangle$ and integrating,

$$\int \langle a|x\rangle \hat{A}\left(x, \frac{\partial}{\partial x}\right)\langle x|a\rangle\,dx = a \int \langle a|x\rangle\langle x|a\rangle\,dx, \quad (12.24)$$

or
$$\langle a|\hat{A}|a\rangle = a\langle a|a\rangle.$$

Now the left-hand side is real, by (12.22), and $\langle a|a\rangle$ is the normalization integral, (12.3), which is real from the definition. Hence a is real.

This result is clearly required for the physical consistency of the Interpretive Postulate, I(i), of § 3.2, that the eigenvalues of an observable operator are the possible results of the corresponding observation.

THEOREM II. *The eigen-vectors belonging to different eigenvalues of an observable operator are orthogonal (have zero overlap).*

Proof. Let a_1 and a_2 be different eigenvalues of $\hat{A}$. Then

$$\hat{A}|a_1\rangle = a_1|a_1\rangle, \quad (12.25)$$

$$\hat{A}|a_2\rangle = a_2|a_2\rangle. \quad (12.26)$$

Hence, from (12.25),

$$\langle a_2|\hat{A}|a_1\rangle = a_1\langle a_2|a_1\rangle. \quad (12.27)$$

From (12.26),

$$\langle a_1|\hat{A}|a_2\rangle = a_2\langle a_1|a_2\rangle. \quad (12.28)$$

Taking the complex conjugate of this equation and using (12.22), (12.8) and Theorem I,

$$\langle a_2|\hat{A}|a_1\rangle = a_2\langle a_2|a_1\rangle. \quad (12.29)$$

Subtracting (12.29) from (12.27) gives

$$(a_1 - a_2)\langle a_2|a_1\rangle = 0, \quad (12.30)$$

and, hence,

$$\langle a_2|a_1\rangle = 0, \quad (12.31)$$

or, equivalently,

$$\int \langle a_2|x\rangle\langle x|a_1\rangle\,dx = 0, \quad (12.32)$$

which completes the proof.

Note that the proof depends on (12.22); the weaker condition (12.21) is not sufficient. The proof has deliberately been given in the shorthand notation. The bewildered reader will find it extremely instructive to go through it, putting in explicitly all the implicit x-dependence and integration.

If the eigenstates of an operator are all normalized, the normalization condition can be combined with the result of Theorem II, to give the orthonormality condition

$$\langle a_n | a_{n'} \rangle = \delta_{nn'} \tag{12.33}$$

where
$$\delta_{nn'} = 1, \qquad n = n',$$
$$= 0, \qquad n \neq n'. \tag{12.34}$$

In terms of the energy eigenstates of the hydrogen atom, this reads

$$\int u_{nlm}^*(r, \theta, \phi) \, u_{n'l'm'}(r, \theta, \phi) \, r^2 \, dr \, d\Omega$$

$$\equiv \int \langle n, l, m | r, \theta, \phi \rangle \langle r, \theta, \phi | n', l', m' \rangle r^2 \, dr \, d\Omega$$

$$= \delta_{nn'} \, \delta_{ll'} \, \delta_{mm'}. \tag{12.35}$$

§ 12.3 The Dirac δ-function

The Dirac δ-function is a generalization of the Kronecker $\delta_{nn'}$ when the discrete variables n, n' are replaced by continuous variables a, and a' say. By definition

$$\delta(a - a') = 0, \qquad a \neq a', \tag{12.36}$$

but
$$\int \delta(a - a') \, da = 1, \tag{12.37}$$

provided the range of integration includes the point $a = a'$. For continuous eigenvalues the orthonormality condition reads

$$\langle a | a' \rangle = \delta(a - a'). \tag{12.38}$$

The simplest representation of the δ-function is

$$\delta(a) = \frac{1}{2\pi} \int_{-\infty}^{\infty} e^{iax} \, dx. \tag{12.39}$$

To see this, consider

$$\phi_g(a) = \frac{1}{2\pi} \int_0^g e^{iax} \, dx = \frac{1}{\pi} \frac{\sin ag}{a}.$$

Now

$$\int_\infty^\infty \phi_g(a) \, da = \frac{1}{\pi} \int_{-\infty}^{\infty} \frac{\sin ag}{a} \, da = 1. \tag{12.40}$$

But

$$\phi_g(0) = \frac{g}{\pi}$$

and

$$\phi_g(\pm \pi/g) = 0,$$

so that in the limit $g \to \infty$ the entire contribution to the integral (12.40) comes from $a = 0$ (see Fig. 12.1), and both conditions (12.36) and (12.37) are satisfied. This establishes (12.39).

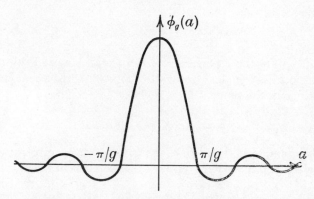

FIG. 12.1. Diagram of the function $\phi_g(a)$. In the limit $g \to \infty$, the peak at the origin becomes infinitely high and narrow, but (12.40) is still valid. So. $\phi_{g \to \infty}(a) = \delta(a)$.

If the de Broglie waves for particles of definite momenta in an infinite volume are normalized according to (12.38) we have (in one dimension)

$$\langle x|p \rangle = c \, e^{ipx/\hbar}$$

where

$$\langle p|p' \rangle = \int \langle p|x \rangle \langle x|p' \rangle dx$$

$$= |c|^2 \int_{-\infty}^{\infty} e^{-i(p-p')x/\hbar} dx$$

$$= \delta(p-p'). \tag{12.41}$$

By comparison of (12.41) with (12.39)

$$|c|^2 = \frac{1}{2\pi\hbar}. \tag{12.42}$$

§ 12.4 Completeness

In addition to the condition (12.22), we further assume that for an operator to represent an observable, its eigenfunctions must form a complete set.† By this we mean that any physical state $\langle x|\psi \rangle$ can be expressed, exactly, as a linear expansion of the eigenfunctions $\langle x|a_n \rangle$ of an operator $\hat{A}$, if $\hat{A}$ represents an observable property of the system. Thus, "completeness" asserts the possibility of making an expansion of any $\langle x|\psi \rangle$ in the form:

$$\langle x|\psi \rangle = \sum_m \langle x|a_m \rangle \, F(a_m), \tag{12.43}$$

$$\langle \psi|x \rangle = \sum_m F^*(a_m) \, \langle a_m|x \rangle, \tag{12.44}$$

where $F(a_m)$ are the coefficients in the expansion, and, by (12.2), (12.44) is just the complex conjugate of (12.43). If the eigenvalues a_m are continuous, the summation is replaced by an integral,

$$\langle x|\psi \rangle = \int \langle x|a \rangle \, F(a) \, da. \tag{12.45}$$

Thus, for example, taking $\hat{A}$ to be the momentum operator, any state function can be expanded in terms of the momentum eigenfunctions.

$$\langle x|\psi \rangle = \int \langle x|p \rangle \, F(p) \, dp,$$

$$= \left(\frac{1}{2\pi\hbar}\right)^{1/2} \int e^{ipx/\hbar} \, F(p) \, dp, \tag{12.46}$$

since
$$\langle x|p \rangle = \left(\frac{1}{2\pi\hbar}\right)^{1/2} e^{ipx/\hbar} \tag{12.47}$$

is the eigenfunction of the operator $\hat{p}$, corresponding to the eigenvalue, p, normalized according to (12.42).

Now substitute (12.43) and (12.44) into the expression for the average of repeated measurements, (12.20).

$$\bar{a}_\psi = \int \left(\sum_n F^*(a_n) \, \langle a_n|x \rangle \right) \hat{A}\left(x, \frac{\partial}{\partial x}\right) \left(\sum_m \langle x|a_m \rangle \, F(a_m)\right) dx$$

$$= \int \left(\sum_n F^*(a_n) \, \langle a_n|x \rangle \right) \left(\sum_m a_m \langle x|a_m \rangle \, F(a_m)\right) dx$$

$$= \sum_n a_n |F(a_n)|^2 \tag{12.48}$$

---

† For matrix operators of finite dimensions, (12.22) implies that they are Hermitian, and the completeness of the eigenvectors follows as a theorem.

The second expression on the right-hand side is obtained by using the eigenvalue equation, (12.11), and the final expression is then a consequence of the orthonormality condition, (12.33).

It follows from (12.48) that the probability of getting a particular result a_n in a single observation, $\hat{A}$, on a system in the state $\langle x|\psi\rangle$, is

$$\mathscr{P}_\psi(a_n) = |F(a_n)|^2. \tag{12.49}$$

(The above argument is exactly analogous to the one given in connection with the spatial distribution, to derive (3.38) in § 3.5.)

In order to find $F(a_n)$, multiply (12.43) by $\langle a_n|x\rangle$ and integrate using the orthonormality condition, (12.33), on the right-hand side:

$$\int \langle a_n|x\rangle \langle x|\psi\rangle \, dx = \int \langle a_n|x\rangle \sum_m \langle x|a_m\rangle F(a_m) \, dx$$
$$= F(a_n).$$

Thus

$$F(a_n) = \langle a_n|\psi\rangle = \int \langle a_n|x\rangle \langle x|\psi\rangle \, dx, \tag{12.50}$$

which is just the overlap integral between the state in question, $\langle x|\psi\rangle$, and the appropriate eigenfunction, $\langle x|a_n\rangle$. The expansion of an arbitrary state is, (12.43),

$$\langle x|\psi\rangle = \sum_{a_n} \langle x|a_n\rangle \langle a_n|\psi\rangle. \tag{12.51}$$

The probability of a result a_n in a measurement of $\hat{A}$ on $|\psi\rangle$ is then, (12.49),

$$\mathscr{P}_\psi(a_n) = |\langle a_n|\psi\rangle|^2, \tag{12.52}$$

which is precisely (3.39), in the Dirac notation.

Note that this is a direct generalization of the standard physical interpretation of a state function, giving the probability density in configuration space, (3.38), which in the Dirac notation reads,

$$\mathscr{P}_\psi(x) = |\langle x|\psi\rangle|^2.$$

The completeness condition is contained in the two equations (12.43) and (12.44). A more formal statement of the completeness condition is that for any set of eigenvectors $|a\rangle$,

$$S_a \ldots |a\rangle\langle a| = \hat{1}, \tag{12.53}$$

where $\qquad S_a = \sum_{a_n},\qquad$ if $a = a_n$ (discrete),

$$= \int \ldots da, \quad \text{if } a \text{ is a continuous variable,}$$

the summation being taken over all allowed physical values in both cases. The meaning of (12.53) is that *any* expression of the form $\langle\phi|\psi\rangle$, can be split into $\langle\phi|$ and $|\psi\rangle$ and the above operation inserted in the middle without changing its numerical value. Thus

$$S_a\langle\phi|a\rangle\langle a|\psi\rangle = \langle\phi|\psi\rangle. \tag{12.54}$$

(12.50) is an example of this, with

$$\phi = a_n,$$
$$a = x.$$

Equation (12.51) is also of this form with

$$\phi = x.$$

This is an extremely useful mnemonic and reduces the construction of the expansion of an arbitrary state $\langle x|\psi\rangle$, and the evaluation of the probabilties, $\mathscr{P}_\psi(a_n)$, to an automatic procedure.

The reader has probably met expansions of the type (12.51) in other contexts. If

$$\hat{A} = \hat{p}, \tag{12.55}$$

the expansion of any given state, $\langle x|\psi\rangle$, is

$$\langle x|\psi\rangle = S_p\langle x|p\rangle\langle p|\psi\rangle.$$
$$= \left(\frac{1}{2\pi\hbar}\right)^{1/2}\int e^{ipx/\hbar}\langle p|\psi\rangle\,dp. \tag{12.56}$$

The "coefficients" in the expansion are

$$\langle p|\psi\rangle = S_x\langle p|x\rangle\langle x|\psi\rangle,$$
$$= \left(\frac{1}{2\pi\hbar}\right)^{1/2}\int e^{-ipx/\hbar}\langle x|\psi\rangle\,dx, \tag{12.57}$$

and the probability of the system having momentum p is

$$\mathscr{P}_\psi(p) = |\langle p|\psi\rangle|^2. \tag{12.58}$$

The two functions $\langle x|\psi\rangle$ and $\langle p|\psi\rangle$, which determine the probability distribution of the particle in configuration space and momentum space, respectively, are the Fourier transforms of each other. A specific calculation of $\langle p|\psi\rangle$, (then called $\phi(p)$,) for the case

$$\langle x|\psi\rangle = \psi(x) = \exp\left[-\frac{x^2}{2\varDelta_x^2}\right] \tag{12.59}$$

is made in § 3.6, in connection with the uncertainty principle.

If the particle is confined to an infinite square well, with potential defined in (3.21), an arbitrary state vector $\langle x|\psi\rangle$ is some arbitrary function in the region $|x| \leqslant a$. This can be expanded, according to (12.51), in terms of the energy eigenfunction, determined by (3.26)–(3.29),

$$\langle x|\psi\rangle = \sum_n \langle x|E_n\rangle\langle E_n|\psi\rangle$$

$$= \sum_{n=1}^{\infty} (a)^{-1/2} \sin\frac{2n\pi x}{2a} \langle E_{2n}|\psi\rangle$$

$$+ \sum_{n=0}^{\infty} (a)^{-1/2} \cos\frac{(2n+1)\pi x}{2a} \langle E_{2n+1}|\psi\rangle, \quad (12.60)$$

which is just the Fourier series expansion. Substituting in (12.50) for the coefficients $\langle E_n|\psi\rangle$, which determine the probability of finding the particle in the various possible energy states, just reproduces the standard formula for determining the coefficients of a Fourier series.

The corresponding expressions for a linear harmonic oscillator are expansions in terms of Hermite polynomials mentioned after (5.26).

It is important that the expansion is made in terms of a uniquely specified set of states. Thus an arbitrary state for an electron in a hydrogen atom is $\langle r,\theta,\phi|\psi\rangle$. This can be expanded in terms of the energy eigenstates, but the eigenvalues of $\hat{l}^2$ and $\hat{l}_z$ must also be specified to give a well-defined set. Thus, by (12.51) in the notation used in (12.35),

$$\langle r,\theta,\phi|\psi\rangle = \sum_{n,l,m} \langle r,\theta,\phi|n,l,m\rangle\langle n,l,m|\psi\rangle, \quad (12.61)$$

where the summation is over all values of n, l and m consistent with the restrictions of § 7.3. The probability of the particle having E, l^2 and $\hat{l}_z$ corresponding to n, l and m is determined by the modulus square of the coefficient

$$\langle n,l,m|\psi\rangle = S_{r,\phi,\theta}\langle n,l,m|r,\theta,\phi\rangle\langle r,\theta,\phi|\psi\rangle,$$

$$= \iint u_{nlm}^*(r,\theta,\phi)\langle r,\theta,\phi|\psi\rangle r^2\,dr\,d\Omega. \quad (12.62)$$

This is a number—the overlap integral—which depends on the given state function $\langle r,\theta,\phi|\psi\rangle$ and the known eigenstate

$$u_{nlm}^*(r,\theta,\phi) \equiv \langle n,l,m|r,\theta,\phi\rangle. \quad (12.63)$$

For any state for a particle in an arbitrary central potential, it is always possible to carry out the expansion of the angular dependence in terms of the eigenstates of angular momentum. Thus

$$\langle r, \theta, \phi | \psi \rangle = \sum_{l,m} \langle \theta, \phi | l, m \rangle \langle l, m, r | \psi \rangle$$

$$= \sum_{l,m} Y_l^m(\theta, \phi) \langle l, m, r | \psi \rangle. \tag{12.64}$$

The "coefficients" are now, by (12.50),

$$\langle l, m, r | \psi \rangle = S_{\theta,\phi} \langle lm | \theta, \phi \rangle \langle r, \theta, \phi | \psi \rangle$$

$$= \int Y_l^m(\theta, \phi)^* \langle r, \theta, \phi | \psi \rangle \, d\Omega. \tag{12.65}$$

Notice that these are functions of r. Thus in the old notation

$$\langle l, m, r | \psi \rangle = \psi_{lm}(r). \tag{12.66}$$

The first term in the expansion (12.64), with coefficient determined by (12.65), is calculated in § 10.4 for the special case of a state vector representing a beam of momentum $\mathbf{p}$, (see (10.30)),

$$\langle r, \theta, \phi | \psi \rangle = \langle r, \theta, \phi | \mathbf{p} \rangle$$

$$= e^{ipr\cos\theta/\hbar}. \tag{12.67}$$

§ 12.5. Operator Methods

(a) The Harmonic Oscillator

These operator techniques can be applied directly to the problem of determining the energy levels of a harmonic oscillator, discussed in Chapter 5. Only the eigenvalue equation (12.12) and the commutation relations are used. It is not necessary to introduce explicit representations of the operators, either as matrices or as derivatives, though the reader may find it easier to regard the following operator algebra as shorthand for matrix manipulations.

The Hamiltonian according to (5.6) is

$$\hat{H} = \frac{\hat{p}^2}{2m} + \tfrac{1}{2}m\omega^2 \hat{x}^2. \tag{12.68}$$

Define

$$\hat{a} = \left(\frac{m}{2}\right)^{1/2} \omega \hat{x} + \frac{i}{(2m)^{1/2}} \hat{p}, \tag{12.69}$$

$$\hat{a}^+ = \left(\frac{m}{2}\right)^{1/2} \omega \hat{x} - \frac{i}{(2m)^{1/2}} \hat{p}. \tag{12.70}$$

These are more general definitions of the operators introduced in Problem 5.4. From the commutation relation for $[\hat{x}, \hat{p}]$, (3.13), it follows by direct substitution that

$$\hat{a}\hat{a}^+ = \hat{H} + \tfrac{1}{2}\hbar\omega, \tag{12.71}$$

$$\hat{a}^+\hat{a} = \hat{H} - \tfrac{1}{2}\hbar\omega, \tag{12.72}$$

and
$$[\hat{a}, \hat{a}^+] = \hbar\omega. \tag{12.73}$$

The eigenstate of energy E_n is $|E_n\rangle$, and

$$\hat{H}|E_n\rangle = E_n|E_n\rangle. \tag{12.74}$$

This can be rewritten, using (12.71) and (12.72), either as

$$\hat{a}\hat{a}^+|E_n\rangle = (E_n + \tfrac{1}{2}\hbar\omega)|E_n\rangle, \tag{12.75}$$

or
$$\hat{a}^+\hat{a}|E_n\rangle = (E_n - \tfrac{1}{2}\hbar\omega)|E_n\rangle. \tag{12.76}$$

These equations correspond directly to (5.11a) and (5.11b). Multiply (12.75) by $\hat{a}^+$:

$$\hat{a}^+\hat{a}\hat{a}^+|E_n\rangle = (E_n + \tfrac{1}{2}\hbar\omega)\,\hat{a}^+|E_n\rangle. \tag{12.77}$$

Then, either
$$\hat{a}^+|E_n\rangle = 0, \tag{12.78}$$

or
$$\hat{a}^+|E_n = |E_{n+1}\rangle, \text{ say,} \tag{12.79}$$

and (12.77) can be written

$$\hat{a}^+ a|E_{n+1}\rangle = [(E_n + \hbar\omega) - \tfrac{1}{2}\hbar\omega]|E_{n+1}\rangle. \tag{12.80}$$

This is (12.76) for $|E_{n+1}\rangle$, provided

$$E_{n+1} = E_n + \hbar\omega. \tag{12.81}$$

Thus, given any eigenvector $|E_n\rangle$, it is possible to generate a new eigenvector $|E_{n+1}\rangle$, by (12.79), with eigenvalue given by (12.81), unless E_n is the highest level, in which case (12.78) is satisfied. But the shape of the potential shows that there is no highest level and the procedure for generating higher levels is always possible.

Similarly, multiplying (12.76) by $\hat{a}$,

$$\hat{a}\hat{a}^+\hat{a}|E_n\rangle = (E_n - \tfrac{1}{2}\hbar\omega)\,\hat{a}|E_n\rangle. \tag{12.82}$$

Now, either
$$\hat{a}|E_n\rangle = 0, \tag{12.83}$$

or
$$\hat{a}|E_n\rangle = |E_{n-1}\rangle, \text{ say.} \tag{12.84}$$

In the latter case (12.82) can be written

$$\hat{a}\hat{a}^+|E_{n-1}\rangle = [(E_n-\hbar\omega)+\tfrac{1}{2}\hbar\omega]|E_{n-1}\rangle. \tag{12.85}$$

This is (12.75) for $|E_{n-1}\rangle$, provided

$$E_{n-1} = E_n-\hbar\omega. \tag{12.86}$$

Thus, given any eigenvector $|E_n\rangle$, it is possible to generate a new eigenvector, $(E_{n-1}\rangle$, by (12.84), unless $|E_n\rangle$ is the ground state, $|E_0\rangle$. In that case (12.83) is satisfied. It then follows, from (12.76), for $n=0$ that

$$E_0-\tfrac{1}{2}\hbar\omega = 0.$$

This determines the ground state energy, and (12.81) then shows that the general level is

$$E_n = (n+\tfrac{1}{2})\hbar\omega, \qquad n = 0,1,2\ldots, \tag{12.87}$$

in agreement with (5.25).

It is clear from (12.79) and (12.84) that $\hat{a}$ and $\hat{a}^+$ are the operators, which, respectively, annihilate and create energy in the system, in units of $\hbar\omega$.

(b) Angular Momentum

The operator techniques employed above for the harmonic oscillator may also be used to establish the possibility of both integer and half-integer values for the angular momentum. This is shown explicitly for the simplest half-integer case in § 8.3. We here prove the general result. To emphasize that we are using a more general definition of angular momentum we denote the operators by $\hat{j}$, rather than $\hat{l}$.

Take the defining property of the angular momentum operators to be the commutation relations stated in (8.31),

$$[\hat{j}_x,\hat{j}_y] = i\hbar\hat{j}_z, \tag{12.88}$$

with two other relations obtained by cyclic permutation of the suffices x, y, z. Define two new operators

$$\hat{j}_+ \equiv \hat{j}_x+i\hat{j}_y, \tag{12.89}$$

$$\hat{j}_- \equiv \hat{j}_x-i\hat{j}_y. \tag{12.90}$$

By direct substitution, it follows from (12.88) that

$$[\hat{j}_z,\hat{j}_+]] = \hbar\hat{j}_+, \tag{12.91}$$

$$[\hat{j}_z,\hat{j}_-] = -\hbar\hat{j}_-, \tag{12.92}$$

and

$$[\hat{j}_+,\hat{j}_-] = 2\hbar\hat{j}_z. \tag{12.93}$$

Since the total angular momentum, $\hat{j}^2$, and $\hat{j}_z$ commute, we can introduce a state $|\beta, m\rangle$ which is simultaneously an eigenstate of both operators. Thus

$$\hat{j}^2|\beta, m\rangle = \hbar^2 \beta |\beta, m\rangle, \tag{12.94}$$

$$\hat{j}_z|\beta, m\rangle = \hbar m |\beta, m\rangle. \tag{12.95}$$

Our problem is to find the possible values of β and m, implied by the commutation relations (12.88). For a given value of the magnitude of the angular momentum, β, it is clear on physical grounds that the possible values of the z-component, specified by m, must lie in a restricted range, with values bounded by m_{max} and m_{min}, say. This remark is important in the subsequent argument.

From (12.95)

$$\hat{j}_+\hat{j}_z|\beta, m\rangle = \hbar m \hat{j}_+|\beta, m\rangle. \tag{12.96}$$

But, by (12.91)

$$\hat{j}_+\hat{j}_z \equiv \hat{j}_z\hat{j}_+ - [\hat{j}_z, \hat{j}_+],$$
$$= \hat{j}_z\hat{j}_+ - \hbar\hat{j}_+. \tag{12.97}$$

Substituting into (12.96), this gives

$$\hat{j}_z\hat{j}_+|\beta, m\rangle = \hbar(m+1)\hat{j}_+|\beta, m\rangle. \tag{12.98}$$

Thus, either

$$\hat{j}_+|\beta, m\rangle = 0, \tag{12.99}$$

or

$$\hat{j}_+|\beta, m\rangle = |\beta, m+1\rangle, \text{ say,} \tag{12.100}$$

and (12.98) can be rewritten

$$\hat{j}_z|\beta, m+1\rangle = \hbar(m+1)|\beta, m+1\rangle. \tag{12.101}$$

Comparing (12.101) with (12.95) it is clear that, given any eigenstate $|\beta, m\rangle$, a new eigenstate $|\beta, m+1\rangle$ can be generated by (12.100), with eigenvalue $m+1$, unless $m = m_{max}$ in which case (12.99) applies. The allowed values of m differ by integers.

By a precisely similar argument applied to $\hat{j}_-\hat{j}_z$ it can be shown that either

$$\hat{j}_-|\beta, m\rangle \quad = 0, \tag{12.102}$$

or

$$\hat{j}_-|\beta, m\rangle \quad = |\beta, m-1\rangle, \tag{12.103}$$

where

$$\hat{j}_z|\beta, m-1\rangle = \hbar(m-1)|\beta, m-1\rangle. \tag{12.104}$$

11

Thus, given any eigenstate $|\beta, m\rangle$, a new eigenstate $|\beta, m-1\rangle$ can be generated by (12.103), with eigenvalue $\hbar(m-1)$, unless $m = m_{\min}$, in which case (12.102) applies.

Now

$$
\begin{aligned}
\hat{\jmath}_-\hat{\jmath}_+ &= (\hat{\jmath}_x - i\hat{\jmath}_y)(\hat{\jmath}_x + i\hat{\jmath}_y), \\
&= \hat{\jmath}_x^2 + \hat{\jmath}_y^2 + i[\hat{\jmath}_x, \hat{\jmath}_y], \\
&= \hat{\jmath}^2 - \hat{\jmath}_z^2 - \hbar\hat{\jmath}_z.
\end{aligned}
\tag{12.105}
$$

If this operates on $|\beta, m_{\max}\rangle$ we get, by (12.99),

$$(\hat{\jmath}^2 - \hat{\jmath}_z^2 - \hbar\hat{\jmath}_z)|\beta, m_{\max}\rangle = \hat{\jmath}_-\hat{\jmath}_+|\beta, m_{\max}\rangle = 0. \tag{12.106}$$

Therefore, by (12.94) and (12.95),

$$(\beta - m_{\max}^2 - m_{\max})|\beta, m_{\max}\rangle = 0,$$

or

$$\beta = m_{\max}(m_{\max} + 1). \tag{12.107}$$

Similarly,

$$\hat{\jmath}_+\hat{\jmath}_- = \hat{\jmath}^2 - \hat{\jmath}_z^2 + \hbar\hat{\jmath}_z. \tag{12.108}$$

Operating with this on $|\beta, m_{\min}\rangle$, and using (12.102), it follows that

$$\beta - m_{\min}^2 + m_{\min} = 0. \tag{12.109}$$

Equating the two expressions for β, (12.107) and (12.109),

$$(m_{\min} + m_{\max})(m_{\min} - m_{\max} - 1) = 0. \tag{12.110}$$

Therefore

$$m_{\min} = -m_{\max} \tag{12.111}$$

Thus the allowed values of m lie symmetrically about the origin. Since the extreme values differ by an integer

$$m_{\max} - m_{\min} = 2j, \ j = 0, \tfrac{1}{2}, 1, \tfrac{3}{2}, \ldots \tag{12.112}$$

Combined with (12.111), this shows that

$$-j \leqslant m \leqslant j, \ (2j+1 \text{ values}) \tag{12.113}$$

and, by (12.107) and (12.112)

$$\beta = j(j+1), \qquad j = 0, \tfrac{1}{2}, 1, \tfrac{3}{2}, \ldots \tag{12.114}$$

Hence, if the commutation relations (12.88) rather than the derivative operators (6.2), are taken as the definitions of angular momentum, both half integer and integer eigenvalues are allowed.

§ 12.6. Summary

(i) A general state function is written

$$\psi(x) = \langle x|\psi\rangle,$$
$$\psi^*(x) = \langle\psi|x\rangle.$$

The normalization is

$$\langle\psi|\psi\rangle \equiv \int \langle\psi|x\rangle\langle x|\psi\rangle \, dx = 1.$$

The overlap of two states $\langle x|\psi\rangle$ and $\langle x|\phi\rangle$ is the generalized scalar product:

$$\langle\phi|\psi\rangle = \int \langle\phi|x\rangle\langle x|\psi\rangle \, dx.$$

If states are orthogonal,

$$\langle\phi|\psi\rangle = 0.$$

(ii) An operator $\hat{A}$ can represent an observation if:
(a) for any $\langle x|\phi\rangle$ and $\langle x|\psi\rangle$,

$$\langle\phi|\hat{A}|\psi\rangle = \langle\psi|\hat{A}|\phi\rangle^*,$$

and

(b) the eigenstates form a complete set (see (iv) below).

The eigenvector corresponding to a_n is $\langle x|a_n\rangle$, so that the eigenvalue equation is

$$\hat{A}\left(x,\frac{\partial}{\partial x}\right)\langle x|a_n\rangle = a_n\langle x|a_n\rangle,$$

or
$$\hat{A}|a_n\rangle = a_n|a_n\rangle.$$

(iii) The eigenstates satisfy the *orthonormality* condition

$$\langle a_n|a_m\rangle = \int \langle a_n|x\rangle\langle x|a_m\rangle \, dx = \delta(a_n, a_m),$$

where the δ symbol denotes $\delta_{a_n a_m}$ or $\delta(a_n - a_m)$ for discrete or continuous a's respectively.

(iv) The eigenstates also satisfy the *completeness* condition, which may be expressed formally as

$$S_a \ldots |a\rangle\langle a| = \hat{1},$$

(see discussion following (12.53)).

(v) These properties of the eigenstates imply that an arbitrary state $\langle x|\psi\rangle$, may be expanded in terms of the eigenstates of $\hat{A}$;

$$\langle x|\psi\rangle = \sum_{a_n} \langle x|a_n\rangle \langle a_n|\psi\rangle,$$

where the coefficients in the expansion are

$$\langle a_n|\psi\rangle = \int \langle a_n|x\rangle \langle x|\psi\rangle \, dx$$

The probability that a single measurement, $\hat{A}$, on a system in the state $|\psi\rangle$, has the result a_n is

$$\mathscr{P}_\psi(a_n) = |\langle a_n|\psi\rangle|^2.$$

All the operators introduced above, to represent the observation of observables, do in fact satisfy the conditions stated under (ii), and their eigenstates have the above properties. Examples of other types of operators are the annihilation and creation operators $(\partial/\partial y \pm y)$, which appear in Chapter 5 in the discussion of the harmonic oscillator and reappear as $\hat{a}$ and $\hat{a}^+$ in § 12.5. They do not represent observables, and do not have the properties listed.

PROBLEMS XII

12.1. The normalized eigenfunction of the ground state, E_0, of a harmonic oscillator is

$$\langle x|E_0\rangle = \frac{\alpha^{1/2}}{\pi^{1/4}} \exp\left[-\tfrac{1}{2}\alpha^2 x^2\right].$$

If the system is in the state

$$\langle x|\psi\rangle = \frac{\sigma^{1/2}}{\pi^{1/4}} \exp\left[-\tfrac{1}{2}\sigma^2 x^2\right],$$

what is the probability that a measurement of the energy gives the result E_0? (For the integral see Problems III, **3.3**.)

12.2. Set up Problems III, **3.4** and **3.5**, using the Dirac notation.

12.3. The eigenfunction of the first excited state of a harmonic oscillator, not normalized, is

$$\langle x|E_1\rangle = x \exp\left[-\tfrac{1}{2}\alpha^2 x^2\right].$$

Find the corresponding eigenfunction in momentum space $\langle p|E_1\rangle$.

CHAPTER 13

EQUATIONS OF MOTION

§ 13.1 The Schrödinger Equation of Motion

We have so far made no mention of time. In particular, we have made no statement about how the observable properties of a quantum system change with the passage of time. It is perhaps a little surprising that we have been able to say so much without introducing an equation of motion. Our main tool has been the eigenvalue equation for the energy—Schrödinger's equation—and a lot of the discussion has been devoted to finding the allowed energy levels of various systems. This question is trivial in classical mechanics. Where statements have been made about the motion of quantum particles in various potentials (Chapter 4), they have been the quantum analogues of the general remarks one can make classically on the basis of energy conservation,

$$\frac{p^2}{2m} + V(x) = E \text{ (const.)}. \tag{13.1}$$

We have not required any generalization of Newton's equation of motion, which is usually stated in terms of force and acceleration. Expressed in terms of physical quantities, which remain important in quantum mechanics, this takes the form

$$\frac{dp}{dt} = \frac{-\partial V}{\partial x}. \tag{13.2}$$

We now develop the quantum mechanical version of this equation.

In Chapter 3 we postulate the commutation relation

$$[\hat{x}, \hat{p}] = i\hbar.$$

When the operator $\hat{x}$ is treated as an ordinary algebraic variable,

$$\hat{x} \to x,$$

the operator $\hat{p}$ is represented by

$$\hat{p} \to -i\hbar \frac{\partial}{\partial x}.$$

151

Now we expect always to treat t as an algebraic variable and, dimensionally, time is to energy as distance is to momentum, in the sense that

$$[\text{time}] \times [\text{energy}] = [\text{distance}] \times [\text{momentum}] = [\hbar].$$

It is thus extremely plausible to postulate that

$$i\hbar \frac{\partial}{\partial t} = \hat{H}. \tag{13.3}$$

(The sign chosen is conventional, but proves to be most convenient when generalizing to relativistic quantum mechanics.) This is an operator equation, which operates on time dependent state vectors, $|\Psi(t)\rangle$. As far as the time dependence of the state vectors is concerned, it has similar physical content to the corresponding equation for $\hat{p}$. However, we already have an alternative expression for the operation of $\hat{H}(x,p)$ on a state in terms of its spatial dependence. Combining these we are led to the equation

$$i\hbar \frac{\partial}{\partial t} \langle x|\Psi(t)\rangle = \hat{H}\left(x, -i\hbar \frac{\partial}{\partial x}\right) \langle x|\Psi(t)\rangle. \tag{13.4}$$

This is the Schrödinger equation of motion. It is a new physical postulate, which in the notation of § 3.3 we should call P(iii). As it has been presented above, it is a conjecture, based on purely formal arguments. Its ultimate justification is that it leads to predictions, which are in agreement with experiment.

Let us consider the form of its solution. Since the operator on the left-hand side depends only on t, and that on the right, only on x, we may look for a particular solution of the form

$$\langle x|\Psi(t)\rangle = u(x)\,f(t). \tag{13.5}$$

Substituting (13.5) into (13.4), and dividing by $u(x)\,f(t)$, gives

$$\frac{i\hbar \dfrac{\partial}{\partial t} f(t)}{f(t)} = \frac{\hat{H}\left(x, -i\hbar \dfrac{\partial}{\partial x}\right) u(x)}{u(x)}. \tag{13.6}$$

Since the left-hand side now depends only on t, and the right only on x, and these variables can be varied independently, each side must in fact be equal to a constant. For reasons which appear immediately, we call this constant E. Thus

$$\frac{i\hbar \dfrac{\partial f(t)}{\partial t}}{f(t)} = E, \tag{13.7}$$

$$\frac{\hat{H}\left(x, -i\hbar\dfrac{\partial}{\partial x}\right)u(x)}{u(x)} = E. \tag{13.8}$$

The second of these equations, (13.8), is just the energy eigenvalue equation, so that

$$E = E_n, \tag{13.9}$$

and

$$u(x) = \langle x|E_n\rangle. \tag{13.10}$$

The solution to (13.7) is then

$$f(t) = e^{-iE_n t/\hbar}. \tag{13.11}$$

A particular solution to (13.4) is thus

$$\langle x|\Psi(t)\rangle_{E_n} = e^{-iE_n t/\hbar}\langle x|E_n\rangle, \tag{13.12}$$

where E_n is any energy eigenvalue.

The general solution is a linear sum of the particular solutions with arbitrary coefficients,

$$\langle x|\Psi(t)\rangle = \sum_n e^{-iE_n t/\hbar}\langle x|E_n\rangle F(E_n). \tag{13.13}$$

We wish to use this equation to determine the development with time of any arbitrary state function, $\langle x|\psi\rangle$. We must thus satisfy the boundary condition that, at $t = 0$,

$$\langle x|\Psi(0)\rangle = \langle x|\psi\rangle. \tag{13.14}$$

Substituting this into (13.13).

$$\langle x|\psi\rangle = \sum_n \langle x|E_n\rangle F(E_n). \tag{13.15}$$

But this is just the expansion of an arbitrary state function in terms of the energy eigenstates discussed in § 12.3. So that, by (12.43) and (12.50),

$$F(E_n) = \langle E_n|\psi\rangle. \tag{13.16}$$

Thus the time dependent solution, which satisfies (13.14), is

$$\langle x|\Psi(t)\rangle = \sum_n e^{-iE_n t/\hbar}\langle x|E_n\rangle\langle E_n|\psi\rangle, \tag{13.17}$$

where again by (12.50),

$$\langle E_n|\psi\rangle = \int \langle E_n|x\rangle \langle x|\psi\rangle \, dx. \tag{13.18}$$

A very simple example is that of a free particle of definite momentum. In this case

$$\langle x|\psi\rangle = \langle x|p\rangle = \left(\frac{1}{2\pi\hbar}\right)^{1/2} e^{ipx/\hbar}. \tag{13.19}$$

The time-dependent solution is

$$\langle x|\Psi(t)\rangle_p = \left(\frac{1}{2\pi\hbar}\right)^{1/2} e^{i(px-Et)/\hbar}, \tag{13.20}$$

where

$$E = \frac{p^2}{2m}.$$

This is the complete de Broglie wave, which is shown in (1.18) to be required to explain electron diffraction. It appears now for the first time as a consequence of the general quantum mechanical formalism, and is direct evidence for the correctness of the postulated equation of motion (13.4).

To illustrate the implications of (13.14) we may again consider the case of a particle in an infinite square well (§ 3.4). Any general state at $t = 0$ can be expanded in terms of the eigenstates. To save algebra the initial state may be specified as a simple superposition of eigenstates. For example

$$\langle x|\psi\rangle = (2)^{-1/2}[\langle x|E_1\rangle + \langle x|E_3\rangle]. \tag{13.21}$$

$$= \left(\frac{1}{2a}\right)^{1/2}\left[\cos\frac{\pi x}{2a} + \cos\frac{3\pi x}{2a}\right]. \tag{13.22}$$

The solution at time t is

$$\langle x|\Psi(t)\rangle = (2)^{-1/2}[\langle x|E_1\rangle e^{-iE_1t/\hbar} + \langle x|E_3\rangle e^{-iE_3t/\hbar}]$$

$$= \left(\frac{1}{2a}\right)^{1/2}\left[\cos\frac{\pi x}{2a} e^{-iE_1t/\hbar} + \cos\frac{3\pi x}{2a} e^{-iE_3t/\hbar}\right], \tag{13.23}$$

where E_n is determined by (3.30). At time t the probability of finding the particle with energy E_n is

$$\mathscr{P}_{\Psi(t)}(E_n) = |\langle E_n|\Psi(t)\rangle|^2$$

$$= \left|\int \langle E_n|x\rangle \langle x|\Psi(t)\rangle \, dx\right|^2,$$

$$= |\langle E_n|\psi\rangle|^2. \tag{13.24}$$

These probabilities do not change with time, and for the state quoted above the particle is equally likely to be in either the first or third level at any time. However, its probability distribution in space is

$$\mathscr{P}_{\Psi(t)}(x) = |\langle x|\Psi(t)\rangle|^2 = \frac{1}{2a}\left[\cos^2\frac{\pi x}{2a} + \cos^2\frac{3\pi x}{2a}\right]$$
$$+ \frac{1}{a}\cos\frac{\pi x}{2a}\cos\frac{3\pi x}{2a}\cos(E_3 - E_1)\,t/\hbar.$$

The spatial distribution thus has a term which does depend on time and, in particular, the probability of finding the particle at the origin oscillates between zero and its starting value,

$$\mathscr{P}_{\Psi(t)}(0) = \frac{1}{a}(1 + \cos(E_3 - E_1)\,t/\hbar). \tag{13.27}$$

§ 13.2 The Heisenberg Equation of Motion

Schrödinger's equation of motion is extremely valuable in developing approximate methods for evaluating scattering cross-sections. We do not consider this here, but consider the relation of this equation to classical mechanics.

Schrödinger's equation is the most natural way to describe changes with time of a quantum mechanical system. In this picture the operators representing observations are regarded as being independent of time. The state vectors represent the observed systems, and as these change with time, so do the results of observations. The average result of repeated observations, $\hat{A}$, made at time t, on an assembly of systems all in the same state $|\Psi(t)\rangle$ is

$$\bar{a}_{\Psi(t)} = \langle\Psi(t)|\hat{A}|\Psi(t)\rangle$$
$$= \int \langle\Psi(t)|x\rangle\,\hat{A}\left(x, \frac{\partial}{\partial x}\right)\langle x|\Psi(t)\rangle\,dx. \tag{13.28}$$

This, of course, is a function of t.

Although this is a natural description, it does not have any simple classical limit, since classically the state vectors play no role. Also classically the operations of observation are not distinguished from the results of observations, so the operators become ordinary algebraic variables (for example $x(t)$ and $p(t)$), and it is these which exhibit the time dependence of the various physical aspects of the system. We must try to cast (13.28) into a form which is more in the spirit of the classical description. The trick is quite simple, particularly if we use the Dirac notation, which does not clutter the argument with any non-essential information.

In terms of state-vectors, the Schrödinger equation of motion, (13.4), is

$$i\hbar \frac{\partial}{\partial t} |\Psi(t)\rangle = \hat{H}|\Psi(t)\rangle. \tag{13.29}$$

This has the formal solution (including the boundary condition (13.14))†,

$$|\Psi(t)\rangle = e^{\ i\hat{H}t/\hbar}|\Psi(0)\rangle = e^{-i\hat{H}t/\hbar}|\psi\rangle. \tag{13.30}$$

Hence (13.28) can be re-written

$$\bar{a}_{\Psi(t)} = \langle\psi|e^{+i\hat{H}t/\hbar}\,\hat{A}\,e^{-i\hat{H}t/\hbar}|\psi\rangle. \tag{13.31}$$

If we define a time-dependent operator

$$\hat{A}(t) = e^{+i\hat{H}t/\hbar}\,\hat{A}\,e^{-i\hat{H}t/\hbar}, \tag{13.32}$$

then the average value of this operator, for a state $|\psi\rangle$ specified at $t = 0$, is

$$\overline{a(t)_\psi} = \langle\psi|\hat{A}(t)|\psi\rangle. \tag{13.33}$$

Equation (13.33) is actually identical with (13.28). All we have done is to change the notation on the right-hand side so that

$$\bar{a}_{\Psi(t)} = \overline{a(t)_\psi}. \tag{13.34}$$

However, the picture is now quite different. The operator $\hat{A}(t)$ represents the making of an observation at time t on a state, which is specified at $t = 0$. This is much closer to the classical description, and one may expect the time dependent operators, $\hat{A}(t)$, to be rather closely related to the corresponding time-dependent classical variables.

Differentiating (13.32), remembering to keep non-commuting operators in their right order,

$$i\hbar \frac{d\hat{A}(t)}{dt} = -\hat{H}\,e^{i\hat{H}t/\hbar}\,\hat{A}\,e^{-i\hat{H}t/\hbar} + e^{i\hat{H}t/\hbar}\,\hat{A}\,e^{-i\hat{H}t/\hbar}\,\hat{H}$$

$$= -\hat{H}\hat{A}(t) + \hat{A}(t)\,\hat{H}.$$

(Note that

$$\hat{H}(t) = e^{i\hat{H}t/\hbar}\,\hat{H}\,e^{-i\hat{H}t/\hbar} = \hat{H},$$

† The meaning of this, in terms of state functions, is

$$\langle x|\Psi(t)\rangle = \exp\left[-i\hat{H}(x, -i\hbar\,\partial/\partial x)\,t/\hbar\right]\langle x|\psi\rangle,$$

$$= \exp\left[-i\hat{H}t/\hbar\right]\sum_n \langle x|E_n\rangle\langle E_n|\psi\rangle,$$

$$= \sum_n \exp\left[-iE_n t/\hbar\right]\langle x|E_n\rangle\langle E_{..}|\psi\rangle,$$

which is a rederivation of (13.17).

so that the "time dependent" energy operator does not, in fact, depend on t.) This is the Heisenberg equation of motion, which may be written in the slightly more compact form

$$i\hbar \frac{d\hat{A}(t)}{dt} = [\hat{A}(t), \hat{H}]. \qquad (13.35)$$

The physical content of the equation is identical with that of the Schrödinger's equation, (13.4).

Apply (13.35) to

$$\hat{A}(t) = \hat{x}(t), \qquad (13.36)$$

for the one-dimensional motion of a particle moving in a potential,

$$\hat{H} = \frac{\hat{p}^2}{2m} + V(\hat{x}).$$

Then

$$i\hbar \frac{d\hat{x}(t)}{dt} = [\hat{x}(t), \hat{H}(t)]. \qquad (13.37)$$

It is easily verified from the definition (13.32) that the commutation relations of the time-dependent operators (Heisenberg operators) is the same as for the corresponding time-independent ones (Schrödinger operators). Hence

$$i\hbar \frac{d\hat{x}(t)}{dt} = \frac{1}{2m} [\hat{x}(t), \hat{p}^2(t)]. \qquad (13.38)$$

But

$$[\hat{x}, \hat{p}^2] \equiv \hat{x}\hat{p}^2 - \hat{p}^2 \hat{x}$$

$$\equiv [\hat{x}, \hat{p}]\hat{p} + \hat{p}[\hat{x}, \hat{p}]$$

$$= 2i\hbar \hat{p}. \qquad (13.39)$$

Therefore

$$\frac{d\hat{x}(t)}{dt} = \frac{\hat{p}(t)}{m}. \qquad (13.40)$$

Similarly, if

$$\hat{A}(t) = \hat{p}(t) \qquad (13.41)$$

$$i\hbar \frac{d\hat{p}(t)}{dt} = [\hat{p}(t), \hat{H}]$$

$$= [\hat{p}, V(\hat{x})]$$

$$= -i\hbar \frac{\partial V(\hat{x})}{\partial \hat{x}}. \qquad (13.42)$$

To obtain the final equality, we have used the result of Problem **3.2** in Chapter 3. Hence

$$\frac{d\hat{p}(t)}{dt} = -\frac{\partial V(\hat{x})}{\partial \hat{x}}. \tag{13.43}$$

These equations may be trivially extended to three dimensions.

Now (13.40) is formally identical with the classical relation between momentum and velocity (rate of change of position). Equation (13.43) is the direct generalization to operators of Newton's Second Law (13.2). This shows that, the quantum equation of motion (13.4)—or equivalently (13.35)—implies that the time-dependent operators, defined by (13.32), satisfy precisely the same equations as the corresponding classical variables. Alternatively, we can multiply (13.40) on the left by $\langle\psi|x\rangle$, on the right by $\langle x|\psi\rangle$ for any state $|\psi\rangle$, and integrate over x. This gives the average value of repeated observations made at time t for a system, which at $t = 0$ is in any arbitrary state $|\psi\rangle$. Thus

$$\frac{d}{dt}\overline{x(t)_\psi} = \frac{\overline{p(t)_\psi}}{m}. \tag{13.44}$$

Similarly, from (13.43),

$$\frac{d}{dt}\overline{p(t)_\psi} = -\frac{\overline{\partial V(x)_\psi}}{\partial x}. \tag{13.45}$$

This shows that these average values satisfy the classical equations of motion.

We have thus completely satisfied the general statement of the correspondence principle, that in the classical limit, quantum mechanics goes over into classical, or Newtonian, mechanics.

§ 13.3 Constants of the Motion. Parity

The equation of motion in the Heisenberg form, (13.35), is not of much practical importance in particular problems in quantum mechanics, since it refers to the operators. This implies a dependence on all the expectation values of the operator for any state $|\psi\rangle$. However, it leads to simple and important conclusions for any observable, $\hat{F}(t)$, which commutes with $\hat{H}$, such as $\hat{j}^2$ or $\hat{j}_z$,

$$[\hat{F}(t), \hat{H}] = 0. \tag{13.46}$$

Then, by (13.35),

$$\frac{d\hat{F}(t)}{dt} = 0. \tag{13.47}$$

Taking the average value for any state,

$$\frac{d}{dt}\overline{\hat{F}(t)_\psi} = \frac{d}{dt}\langle \Psi(t)|\hat{F}|\Psi(t)\rangle = 0, \qquad (13.48)$$

showing that this does not change with time. If the system is in an eigenstate of $\hat{F}$ at time $t = 0$, the state will be an eigenstate at any subsequent time, since the operator does not change with time.

The operators, $\hat{F}(t)$, satisfying (13.46) are called *constants of the motion*. They are the generalization of the conserved quantities of classical mechanics. For a free particle, or any closed system, the total momentum operator is a constant of the motion. As shown in Chapter 8, for a particle in any central potential, the total angular momentum and each separate component all commute with the Hamiltonian and, from the above argument, are also constants of the motion. ($\hat{j}^2$ and $\hat{j}_z$ may have constant definite values, $\hat{j}_x$ and $\hat{j}_y$ then have only constant average values.)

We may also introduce as constants of the motion some "observable" properties of a quantum system which have no classical analogue.

Consider a system with a Hamiltonian which is unchanged by reflection of the co-ordinates,

$$\hat{H}(\mathbf{r}) = \hat{H}(-\mathbf{r}). \qquad (13.49)$$

Introduce the reflection operator, $\hat{P}$, which by definition has the property that for any state

$$\hat{P}\langle \mathbf{r}|\psi\rangle = \langle -\mathbf{r}|\psi\rangle. \qquad (13.50)$$

Thus the effect of $\hat{P}$ on any state is to turn it into the corresponding state in the reflected co-ordinate system. At first sight this would appear to be an operator somewhat similar to the annihilation and creation operators introduced at the end of Chapter 5. However, it has real eigenvalues, ± 1, corresponding to state functions which are odd or even under reflection. According to § 4.2 the eigenfunctions of $\hat{H}$ can be expressed as states, which have definite reflection properties (odd or even). Since these form a complete set, it is possible to make up a complete and uniquely defined set of states, each of which is an eigenstate of the operator $\hat{P}$. Thus $\hat{P}$ satisfies all the requirements of Chapter 12 to be regarded as an observable operator. Its eigenvalue, ± 1, is the parity of the state, and the operator may be considered to represent the operation of observing the parity.

Now, for *any* state $|\psi\rangle$,

$$\hat{P}\hat{H}(\mathbf{r})\langle\mathbf{r}|\psi\rangle = \hat{H}(-\mathbf{r})\langle-\mathbf{r}|\psi\rangle,$$
$$= \hat{H}(\mathbf{r})\,\hat{P}\langle\mathbf{r}|\psi\rangle,$$

where we have used (13.49). Therefore, we have the operator equation

$$\hat{P}\hat{H}(\mathbf{r}) = \hat{H}(\mathbf{r})\,\hat{P}. \tag{13.51}$$

But this is just the condition that, by (13.35),

$$\frac{d\hat{P}}{dt} = 0, \tag{13.52}$$

so that $\hat{P}$ is a constant of the motion. The parity of any state remains unchanged throughout its development in time provided only that (13.49) is satisfied. This is a new type of conservation law, which plays an important role in quantum mechanics, and has no analogue in classical physics.

Of the four basic interactions of Table 11.2 the first three satisfy (13.51), but the "weak" nuclear interactions, of which a typical example is the β-decay of a neutron,

$$n \to p + e^- + \bar{\nu},$$

are not unchanged by reflection, and under their influence parity is not conserved. This means that such decays look essentially different when viewed in a reflected co-ordinate system—or in a mirror. A classical picture of this is given by comparing a cannon and a rifle. A cannon ball is shot out with no spin, and looks essentially the same in a reflected system. However, a rifle bullet emerges spinning in a definite sense, say, right-handed about its line of flight. In a mirror this will appear left-handed—an essentially distinguishable situation. In a precisely analogous fashion, in the above decay, the $\bar{\nu}$ always emerges with its spin right-handed about its line of flight, and the neutron, which may be regarded in its decay as a neutrino gun, behaves like a rifle—not a cannon.

§ 13.4 Conservation Laws and Invariance

Given any operator $\hat{F}$, one can define the *Hermitian conjugate* operator $\hat{F}^+$ by the requirement that its expectation values for any states $|\phi\rangle$ and $|\psi\rangle$ are given by

$$\langle\phi|\hat{F}^+|\psi\rangle = \langle\psi|\hat{F}|\phi\rangle^*. \tag{13.53}$$

If $\hat{F}$ is a matrix the condition on its elements reads

$$\langle i|\hat{F}^+|j\rangle = \langle j|\hat{F}|i\rangle^*. \tag{13.54}$$

In terms of this definition, the condition (12.22) for an operator to represent an observation (or equivalently, to be an observable) is

$$\hat{F} = \hat{F}^+. \tag{13.55}$$

Such an operator is said to be *Hermitian*.

If an operator $\hat{U}$ satisfies the relation

$$\hat{U}\hat{U}^+ = \hat{1}, \tag{13.56}$$

it is said to be *unitary*. The condition applies to either differential or matrix operators. If $\hat{U}$ is a matrix the multiplication of $\hat{U}$ and $\hat{U}^+$ is as defined in (8.21).

Given any Hermitian operator $\hat{F}$, then one can construct the operator

$$\hat{U} = \exp[i\epsilon\hat{F})$$

$$\equiv \hat{1} + \frac{(i\epsilon\hat{F})}{1!} + \frac{(i\epsilon\hat{F})^2}{2!} + \cdots \tag{13.57}$$

where ϵ is a real number. Since

$$\hat{U}^+ = \exp[-i\epsilon\hat{F}^+] = \exp[-i\epsilon\hat{F}], \tag{13.58}$$

it follows that

$$\hat{U}\hat{U}^+ = \exp[i\epsilon\hat{F}]\exp[-i\epsilon\hat{F}] = \hat{1}, \tag{13.59}$$

and hence that $\hat{U}$ is unitary, as we have already anticipated by the notation. If ϵ is infinitesimal, we need only take the first two terms in the expansion (13.57), so that for an infinitesimal unitary operator

$$\hat{U} = \hat{1} + i\epsilon\hat{F}, \tag{13.60}$$

and

$$\hat{U}^+ = \hat{1} - i\epsilon\hat{F}.$$

In this way a Unitary operator can be related to any Hermitian operator and *vice-versa*. We have already seen that Hermitian operators $\hat{F}$ can be interpreted as physical observables. We now give a physical interpretation to unitary operators.

Let $|\psi\rangle$ be any state. By operating with the unitary operators $\hat{U}$ we can form a new (transformed) state

$$|\psi^u\rangle \equiv \hat{U}|\psi\rangle, \tag{13.61}$$

with components

$$\langle x|\psi^u\rangle = \langle x|\hat{U}|\psi\rangle.$$

Then by the definitions (12.2) and (13.54),

$$\langle\psi^u|x\rangle = \langle x|\psi^u\rangle^* = \langle x|\hat{U}|\psi\rangle^* = \langle\psi|\hat{U}^+|x\rangle, \qquad (13.62)$$

so that

$$\langle\psi^u| = \langle\psi|\hat{U}^+. \qquad (13.63)$$

Thus if $\langle\psi|$ is normalized, then

$$\langle\psi^u|\psi^u\rangle = \langle\psi|\hat{U}^+ U|\psi\rangle = \langle\psi|\psi\rangle = 1, \qquad (13.64)$$

so that $|\psi^u\rangle$ is also normalized. By the same argument, if $|a_n\rangle$ is a complete set of orthonormal eigenstates (see § 12.6), then

$$\hat{U}|a_n\rangle \equiv |a_n^u\rangle \qquad (13.65)$$

is also a complete, orthonormal set. Thus unitary operators enable one to make transformations from one description of a system to another physically equivalent description. For example $|a_n\rangle$ might be the energy eigenstates of a system specified according to a certain set of coordinates. Then, by the appropriate choice of U, $|a_n^u\rangle$ can be the states of the same system with reference to new coordinates with the origin displaced, or the orientation changed.

In general, if $\epsilon\hbar$ is the displacement of the origin of a (cartesian) coordinate, and $\hat{F}$ is the operator representing the momentum corresponding to the coordinate, then $\hat{U}$ defined by (13.57) is the unitary operator which transforms the old states into those with respect to the new coordinates. A simple example of this is the displacement a of the x coordinate of a linear system, such as a one-dimensional harmonic oscillator. Then $\hat{F}$ is the momentum $\hat{p}$, and the operator which induces a displacement by **a** of the origin of coordinates is

$$\hat{U}^a = \exp[i\hat{p}a/\hbar]. \qquad (13.66)$$

Working in the Schrödinger representation the new states $|\psi^a\rangle$ are expressed in terms of the old states $|\psi\rangle$ by

$$\langle x|\psi^a\rangle = \exp\left[\frac{ia}{\hbar}\left(-i\hbar\frac{\partial}{\partial x}\right)\right]\langle x|\psi\rangle,$$

$$= \left[1 + \frac{1}{1!}\left(a\frac{\partial}{\partial x}\right) + \frac{1}{2!}\left(a\frac{\partial}{\partial x}\right)^2 + \ldots\right]\langle x|\psi\rangle$$

$$= \langle x+a|\psi\rangle. \qquad (13.67)$$

The last equality follows since the penultimate expression is just the Taylor expansion of the state function $\langle x+a|\psi\rangle$ in terms of the original state function $\langle x|\psi\rangle$.

Consider an infinitesimal operator $\hat{U}$, of this general type, representing an infinitesimal transformation of a coordinate used to describe a physical system. The system is said to be invariant with respect to these transformations if the expectation value of its Hamiltonian for any states $|\phi\rangle$ and $|\psi\rangle$ is unchanged by the transformation;

$$\langle\phi|\hat{H}|\psi\rangle = \langle\phi^u|\hat{H}|\psi^u\rangle. \tag{13.68}$$

Thus

$$\langle\phi|\hat{H}|\psi\rangle = \langle\phi|\hat{U}^+\hat{H}U|\psi\rangle,$$
$$= \langle\phi|(\hat{1}-i\epsilon\hat{F})\,\hat{H}(1+i\epsilon\hat{F})|\psi\rangle,$$
$$= \langle\phi|\hat{H}|\psi\rangle$$
$$\quad -(i\epsilon)\langle\phi|\hat{F}\hat{H}-\hat{H}\hat{F}|\psi\rangle, \tag{13.69}$$

where, in the final equality we have neglected a term of order ϵ^2. Since this has to be true for any $|\phi\rangle$ and $|\psi\rangle$, we can drop the states and write just the operator condition for invariance;

$$\hat{F}\hat{H}-\hat{H}\hat{F} \equiv [\hat{F},\hat{H}] = 0. \tag{13.70}$$

But by (13.46) this is just the condition that $\hat{F}$ is a constant of the motion—or that $\langle\hat{F}\rangle$ is conserved. Since the argument is clearly reversible we have an important relation between invariance and conservation laws. *The necessary and sufficient condition for a particular momentum to be conserved is that the Hamiltonian be invariant for displacements with respect to the corresponding coordinate.*

As an example the matrix elements of the Hamiltonian

$$\hat{H} = \hat{p}^2/2m$$

for a free particle are invariant for displacements of the coordinate and the linear momentum of the system is conserved. However this invariance does not hold for a particle moving in a harmonic oscillator potential centred on some fixed point, since the potential defines a natural coordinate origin. For such a system momentum is not conserved.

This result is not in fact restricted to transformations of the space–time coordinates and the corresponding linear momentum, but for any pair of complementary observables. The generalization is used in the next chapter.

§ 13.5 Summary

To the two physical postulates of § 3.3 (the Correspondence Principle and the Complementarity Principle), we have added a

12

third—the Equation of Motion for a quantum system. This may be written either in the Schrödinger form

$$i\hbar\frac{\partial}{\partial t}|\Psi(t)\rangle = \hat{H}|\Psi(t)\rangle,$$

or in the Heisenberg form

$$i\hbar\frac{d\hat{A}(t)}{dt} = [\hat{A}(t), \hat{H}].$$

The former is most suitable for quantum calculations. The latter is more closely related to classical theory, and has been shown to imply that classical mechanics is indeed the classical limit ($\hbar\to0$) of quantum mechanics.

The most important application of Heisenberg's equation is to those observables $\hat{F}(t)$ for which the commutator with $\hat{H}$ vanishes. They represent observables which do not change with time and are thus conserved. These conserved (Hermitian) observables can be related to unitary operators which describe transformations of coordinates, with respect to which the Hamiltonian is invariant— such as displacements and rotations. Thus invariance with respect to displacement of linear coordinates is related to the conservation of linear momentum, and invariance with respect to rotations (displacements of angular coordinates) is related to the conservation of angular momentum.

PROBLEMS XIII

13.1. Show that the Heisenberg equations for the operators $\hat{x}(t)$ and $\hat{p}(t)$ of a harmonic oscillator are formally identical with the classical equations of motion.

13.2. In a system of two identical particles we may introduce an operator $\hat{P}_{12}$, which interchanges particles 1 and 2. It has eigenvalues ± 1 for eigenstates which are symmetric or anti-symmetric for the exchange. What is the condition on the Hamiltonian, $\hat{H}$, which ensures that $\hat{P}_{12}$ is a constant of the motion?

CHAPTER 14

THE GOLDEN RULE

§ 14.1 Time-dependent Perturbation Theory

In the study of fluid flow under idealized stable conditions, two possible approaches are possible, associated respectively with the names of Euler and Lagrange. In the Eulerian method one looks at the system as a whole in terms of the density and current at fixed points in space. Time does not appear explicitly since, although the fluid is flowing, the currents and densities at fixed points do not change with time in a steady situation. Alternatively, one may concentrate attention on a particular element of actual fluid, following its motion through the system. In this Lagrangian view, even in a steady state situation, time plays a crucial role, since the position of the test volume element is traced as a function of time. So far, dynamical problems in quantum mechanics have been treated from the overall, time-independent, Eulerian viewpoint. Thus, in § 4.1, the effect of a potential step on a beam of particles is discussed in terms of the transmitted and reflected beams which are induced. In § 10.3 and 10.4 the scattering of a beam of particles by a potential is considered from the same point of view. The time-dependent Schrödinger equation makes it possible to consider dynamical problems of scattering and decay in a manner which is much closer to that of Lagrange. The system starts in a certain state and from the time development of this state as given by the Schrödinger equation (13.29) it is possible to calculate the probability per unit time for finding it at some later instant in some other state. The problem cannot be solved in general, but if the interaction causing the transitions from one state to the other is small, it is possible to obtain an approximate solution in ascending powers of the strength of the interaction potential. This is known as time-dependent perturbation theory. The theory is extremely general and we will therefore work in the general Dirac notation. However, to fix one's ideas, it is useful to think in terms of a particle being scattered by a fixed potential—a problem which has already been considered from the Eulerian viewpoint in Chapter 10.

Consider any system in which the total energy operator—the Hamiltonian—can be split into two parts

$$\hat{H} = \hat{H}_0 + \hat{V}, \tag{14.1}$$

in which $\hat{H}_0$ has a set of eigenstates which can be found exactly and defines the free system. Then $\hat{V}$ is by definition the interaction potential. For the simple scattering problem,

$$\hat{H} = \frac{\hat{p}^2}{2m} + V(\hat{r}), \tag{14.2}$$

and clearly the free system is defined by the kinetic energy term

$$\hat{H}_0 = \frac{\hat{p}^2}{2m}. \tag{14.3}$$

Let the eigenstates of $\hat{H}_0$ of energy E_n be written

$$|E_n, \alpha\rangle \equiv |n\rangle \tag{14.4}$$

where α is shorthand for the extra information required to determine a unique state. For a free particle this could be the two independent components of a unit vector defining the direction of the momentum. Alternatively, one could define the state by the momentum vector:

$$|n\rangle \equiv |\mathbf{p}_n\rangle. \tag{14.5}$$

The states $|n\rangle$ form a complete orthonormal set. Although they would normally be continuous, for the general derivation we can treat them as though they were discrete and write the orthonormality condition

$$\langle n|m\rangle = \delta_{nm}. \tag{14.6}$$

The time dependent free states are then, by (13.13)

$$|n(t)\rangle = |n\rangle\, e^{-iE_n t/\hbar}. \tag{14.7}$$

Let us suppose that the actual physical system is in a time-dependent state

$$|\Psi_i(t)\rangle,$$

which must satisfy the Schrödinger equation of motion (13.29),

$$i\hbar \frac{\partial}{\partial t}\bigg| \Psi_i(t)\rangle = |\hat{H}_0 + \hat{V}|\Psi_i(t)\rangle. \tag{14.8}$$

The suffix i indicates that at an initial time which we take to be

$$t = -T/2, \tag{14.9}$$

the system is in an eigenstate of the free system (see (14.7)):

$$|\Psi_i(-T/2)\rangle = |i\rangle\, e^{iE_i T/2\hbar}. \tag{14.10}$$

For a particle scattering in a potential

$$|i\rangle = |\mathbf{p}_i\rangle \tag{14.11}$$

which defines the energy and direction of the initial beam, before scattering has taken place.

It is convenient to look for a solution to (14.8) as an expansion in the time-dependent eigenstates of the free system:

$$|\Psi_i(t)\rangle = \sum_n |n\rangle\, e^{-iE_n t/\hbar}\, a_{ni}(t) \tag{14.12}$$

If there were no interaction energy ($\hat{V} = 0$) this would be the general form of the time-dependent solution (13.13) with the coefficients a_{ni} [corresponding to $F(E_n)$] constant. However, the presence of $\hat{V}$ in the Hamiltonian induces a time dependence in these coefficients. It is evident that the boundary condition (14.10) is

$$a_{ni}(-T/2) = \delta_{ni}. \tag{14.13}$$

Consider some final state denoted by f which for a particle scattering in a potential is specified by some final momentum $\mathbf{p}_f$, so that

$$|f\rangle = |\mathbf{p}_f\rangle. \tag{14.14}$$

The probability of finding the system at time t in the state $|f\rangle$ is

$$w(t) = |\langle f|\Psi_i(t)\rangle|^2 = |a_{fi}(t)|^2. \tag{14.15}$$

If we consider a total time interval T from the time the system is set up in the initial state $|i\rangle$, we see that the probability per unit time for the system to make a transition to a final state $|f\rangle$ is

$$``w_{fi}" = \frac{w(T/2)}{T} = \frac{|a_{fi}(T/2)|^2}{T}. \tag{14.16}$$

Thus the coefficients $a_{fi}(t)$ in the expansion (14.12) are simply related to the transition rates we wish to calculate and are known as transition amplitudes.

To derive an equation for these amplitudes, substitute (14.12) into (14.8);

$$i\hbar\frac{\partial}{\partial t}\left[\sum_n |n\rangle\, e^{-iE_n t/\hbar}\, a_{ni}(t)\right] = \sum_n |\hat{H}_0 + \hat{V}||n\rangle\, e^{-iE_n t/\hbar}\, a_{ni}(t). \tag{14.17}$$

On the right-hand side we may use the fact that the states are eigenstates of the free system, so that

$$H_0|n\rangle = E_n|n\rangle. \tag{14.18}$$

This gives rise to a series of terms which cancel with those on the left-hand side coming from the operation of the time derivative on the exponential factors. The remaining terms are:

$$\sum_n i\hbar |n\rangle e^{-iE_n t/\hbar} \frac{d}{dt} a_{ni}(t) = \sum_n \hat{V} |n\rangle e^{-iE_n t/\hbar} a_{ni}(t). \qquad (14.19)$$

Now multiply on the left by $\langle f|$ and use the orthonormality condition

$$\langle f|n\rangle = \delta_{fn}. \qquad (14.20)$$

This removes the summation on the left-hand side leaving a single term multiplied by an exponential factor. After rearranging, the equation can be written

$$\frac{d}{dt} a_{fi}(t) = (i\hbar)^{-1} \sum_n \langle f| \hat{V} |n\rangle e^{i(E_f - E_n)t/\hbar} a_{ni}(t), \qquad (14.21)$$

which must be solved subject to the boundary condition (14.13). Both the boundary condition and the differential equation can be combined into the single integral equation

$$a_{fi}(t) = \delta_{fi} + (i\hbar)^{-1} \int_{-T/2}^{t} \sum_n \langle f| \hat{V} |n\rangle e^{i(E_f - E_n)t'/\hbar} a_{ni}(t') \, dt'. \qquad (14.22)$$

To see this, firstly note that if

$$t = -T/2, \qquad (14.23)$$

the range of integration vanishes, so the equation (14.22) simply reduces to the required boundary condition (14.13). Secondly, the derivative with respect to time of (14.22) is just the equation (14.21). This is an important equation for the transition amplitudes $a_{fi}(t)$ since it is exact, no approximations having been made so far. It is, however, not solvable in closed form.

A simple approximate solution can be obtained if it is assumed that we can expand in powers of the interaction energy $\hat{V}$ (or more strictly its matrix element $\langle f| \hat{V} |n\rangle$). Thus the zeroth approximation is to ignore the integral altogether, which gives

$$a_{fi}(t) = \delta_{fi}. \qquad (14.24)$$

This is an algebraic statement of the obvious physical fact that if the interaction energy is neglected the system remains for all time in the initial state (see (14.7) and (14.10))

$$|\Psi_i(t)\rangle = |i(t)\rangle. \qquad (14.25)$$

To get the next approximation, substitute (14.24) into the integrand in (14.22). The summation becomes trivial and the equation can be evaluated at $t = T/2$ to give

$$a_{fi}(T/2) = \delta_{fi} + (i\hbar)^{-1}\langle f|\hat{V}|i\rangle \int_{-T/2}^{T/2} e^{i(E_f-E_i)t/\hbar}\,dt. \qquad (14.26)$$

Let us suppose that the final state is not identical with the initial state so that

$$\delta_{fi} = 0, \qquad (14.27)$$

then (14.26) can be used in (14.16). It is convenient to take the limit of the time interval going to infinity. Thus:

$$"w_{fi}" = (\hbar)^{-2}|\langle f|\hat{V}|i\rangle|^2$$

$$\times \lim_{T\to\infty}\left[\hbar\int_{-T/2}^{T/2} e^{i(E_f-E_i)t/\hbar}\frac{dt}{\hbar}\right]\left[\frac{1}{T}\int_{-T/2}^{T/2} e^{i(E_f-E_i)t/\hbar}\,dt\right]. \qquad (14.28)$$

The factor in the first square bracket is (see (12.39))

$$2\pi\hbar\,\delta(E_f - E_i). \qquad (14.29)$$

Since this vanishes unless

$$E_f = E_i \qquad (14.30)$$

we may equate the exponent to zero in the second integral, in which case the whole of the second square bracket reduces to unity. We are thus led to an expression for the transition rate

$$"w_{fi}" = \frac{2\pi}{\hbar}\langle f|\hat{V}|i\rangle\,\delta(E_f - E_i). \qquad (14.31)$$

The δ-function ensures that the rate is zero unless energy is conserved, but the expression is somewhat formal as a consequence of the infinite time interval which we have considered and the fact that we have calculated the rate into a precisely defined final state. It was in anticipation of this that quotation marks were introduced in (14.16). In practice, there is normally at least one free particle in the final state which can be required to have an energy specified within narrow limits and to be travelling in a given direction defined by an infinitesimal solid angle. Equivalently, the particle momentum can be determined by the detection apparatus to lie within the interval between $\mathbf{p}_f$ and $\mathbf{p}_f + d\mathbf{p}_f$. Let

$$\rho(p_f)\,d^3p_f = \rho(p_f)\,p_f^2\,dp_f\,d\Omega_f \qquad (14.32)$$

be the number of quantum states in this interval. We refer to these as relevant states. The rate for a transition into a relevant final state is obtained by multiplying (14.31) by (14.32). (If there are several final particles there may be several such factors—see (14.97) below.) Define

$$\rho(E_f) \equiv \delta(E_i - E_f)\, \rho(p_f)\, d^3 p_f \tag{14.33}$$

which is evidently the density of relevant final states, compatible with energy conservation. Thus, multiplying (14.31) by (14.32), the realistic transition rate into a relevant state is

$$dw_{fi} = \text{``}w_{fi}\text{''}\, \rho(p_f)\, d^3 p_f \tag{14.34}$$

or, using (14.33),

$$\boxed{dw_{fi} = \frac{2\pi}{\hbar}\, |\langle f|V|i\rangle|^2\, \rho(E_f).} \tag{14.35}$$

We have written the transition rate as a differential because $\rho(E_f)$ is a differential quantity leading to a differential decay rate or cross section. The total rate or cross section is obtained by summing over all possible relevant states, which involves integrating over magnitudes of momenta not already determined by energy conservation and all solid angles in the final state.

Formula (14.35) has been applied to such a wide variety of quantum phenomena that it was called the Golden Rule by Enrico Fermi. We continue with the specific application to the scattering of a single particle in a fixed potential and then discuss radiative transitions and decay rates.

§ 14.2 Potential Scattering

We wish to calculate the differential cross section for the scattering of a particle in a potential. This can be obtained from the rate of transition of the system from an initial momentum state $\mathbf{p}_i$ to a final state $\mathbf{p}_f$.

Let us consider first the factor which describes the density of relevant final states. For a free particle confined to one dimensional motion in the range

$$0 \leqslant x \leqslant L, \tag{14.36}$$

the normalized eigenstate of momentum is (see 3.33)

$$\langle x|p_n\rangle = L^{-1/2}\, e^{ip_n x/\hbar} \tag{14.37}$$

where, by (2.11),

$$p_n = \left(\frac{2\pi\hbar}{L}\right) n. \tag{14.38}$$

Thus the number of states in the momentum range $p \to p + dp$ is

$$dn = \frac{L}{2\pi\hbar} dp. \tag{14.39}$$

In three dimensions the corresponding normalized eigenstate is

$$\langle \mathbf{x} | \mathbf{p}_n \rangle = L^{-3/2} e^{i\mathbf{p}_n \cdot \mathbf{x}/\hbar} \tag{14.40}$$

and the number of states in a momentum interval is then, by the same argument,

$$\rho(p) \, d^3 p = \left(\frac{L}{2\pi\hbar}\right)^3 d^3 p = \left(\frac{L}{2\pi\hbar}\right)^3 p^2 \, dp \, d\Omega. \tag{14.41}$$

We remark in passing that $L^3 d^3 p$ is a "volume" in phase-space and we have established the result, important in statistical mechanics, that there is one quantum state per "volume"

$$(2\pi\hbar)^3 = h^3. \tag{14.42}$$

For scattering into a solid angle $d\Omega_f$ the density of relevant final states, by (14.33), is

$$\rho(E_f) = \delta(E_i - E_f) \, \rho(p_f) \, d^3 p_f$$
$$= \left(\frac{L}{2\pi\hbar}\right)^3 p_f^2 \frac{dp_f}{dE_f} d\Omega_f \delta(E_i - E_f) \, dE_f. \tag{14.43}$$

Integrating over any range of final energy which allows for energy conservation, simply removes the δ-function which, however, requires that

$$E_i = E_f = \frac{p_f^2}{2m}, \tag{14.44}$$

so

$$\frac{dE_f}{dp_f} = \frac{p_f}{m}. \tag{14.45}$$

Thus, substituting into (14.43),

$$\rho(E_f) = \left(\frac{L}{2\pi\hbar}\right)^3 m p_f \, d\Omega_f. \tag{14.46}$$

In this context the volume L^3 in which the states are normalized may be thought of as very large but small enough to exclude the apparatus required to produce the initial beam and detect the scattered particle.

It should not affect the result and it is a minor check on our argument that L^3 does not appear in the final expression for the differential cross section, (see (14.51) below).

We are now in a position to write an explicit formula for the transition rate. The differential cross-section is the transition rate per unit flux. This is discussed in § 10.1 and § 10.3. For the normalization we are now using, the flux in the initial state is

$$\text{Flux} = \rho v = \frac{1}{L^3} \frac{p_i}{m}. \tag{14.47}$$

Combining the Golden Rule (14.35) with (14.46) for the density of final states and (14.47) for the initial flux, the expression for the differential cross-section for scattering into the solid angle $d\Omega$ is (see § 10.3)

$$\sigma(\theta, \phi) = \frac{d\sigma}{d\Omega} = \frac{1}{\text{Flux}} \frac{dw_{fi}}{d\Omega}$$
$$= \left(\frac{mL^3}{p_i}\right)\left(\frac{2\pi}{\hbar}\right) |\langle \mathbf{p}_f | \hat{V} | \mathbf{p}_i \rangle|^2 \left(\frac{L}{2\pi\hbar}\right)^3 mp_f. \tag{14.48}$$

This is a long calculation of which there now remains only the final step of evaluating the matrix element of the potential. This can be unscrambled from the Dirac notation along the lines of (12.22), remembering that we now have a realistic problem in three dimension. Thus for a central, spherically symmetric, potential

$$\langle \mathbf{p}_f | V | \mathbf{p}_i \rangle$$
$$= \int \langle \mathbf{p}_f | \mathbf{r} \rangle V(r) \langle \mathbf{r} | \mathbf{p}_i \rangle \, d^3 r$$
$$= \frac{1}{L^3} \int e^{-i\mathbf{p}_f \cdot \mathbf{r}/\hbar} V(r) \, e^{i\mathbf{p}_i \cdot \mathbf{r}/\hbar} d^3 r$$
$$= \frac{1}{L^3} \int e^{-i\mathbf{K} \cdot \mathbf{r}} V(r) \, d^3 r = \frac{1}{L^3} \tilde{V}(K), \tag{14.49}$$

where

$$\mathbf{K}\hbar = (\mathbf{p}_f - \mathbf{p}_i), \tag{14.50}$$

is the momentum transferred to the particle in the transition from the initial to the final state. Substituting (14.49) into (14.48) we have

$$\frac{d\sigma}{d\Omega} = \left(\frac{m}{2\pi\hbar^2}\right)^2 \frac{p_f}{p_i} |\tilde{V}(K)|^2. \tag{14.51}$$

For elastic scattering the magnitudes of the initial and final momenta must be the same so the explicit dependence on momentum cancels to unity. However, these factors appear respectively from the initial flux and the density of final states, independently of the dynamics of the problem which is contained in $\tilde{V}(K)$. In more general potentials which can change the nature of the scattered particle as well as its direction, this factor is significant and may be dominantly important. From now on we will put

$$p_f = p_i = p. \tag{14.52}$$

By comparison of (14.51) with (10.26) it can be seen that, apart from the numerical factor involving m and $\hbar$, $\tilde{V}(K)$ is the scattering amplitude $f(\theta)$, produced by the potential $V(r)$ and, from (14.49), that it is the Fourier transform of the potential with respect to the momentum transfer. This Fourier transform is a scalar quantity which can only depend on the scalars which can be constructed from $\mathbf{p}_i$ and $\mathbf{p}_f$, namely

$$\mathbf{p}_i^2 = \mathbf{p}_f^2 = 2mE \tag{14.53}$$

or

$$\mathbf{p}_i \cdot \mathbf{p}_f = p^2 \cos \theta \tag{14.54}$$

where θ is the angle through which the particle is scattered. Thus the differential cross section is emerging as a function of the energy and scattering angle—as, of course, it must.

If the range of the potential is R, the time that the particle spends in the potential is

$$\tau = \frac{R}{v} = \frac{Rm}{p}. \tag{14.55}$$

The approximation we have made of expanding in powers of $\langle V \rangle$ is valid if the product of the transit time and scattering potential is small compared with $\hbar$; that is

$$\tau \langle V \rangle < \hbar,$$

or

$$p > \frac{Rm \langle V \rangle}{\hbar}. \tag{14.56}$$

The theory is thus applicable for large momenta or equivalently at high energy. It thus complements the analysis in terms of phase shifts which is useful at low energies, when only a limited number of angular momentum states are involved.

To evaluate the Fourier transform of the potential introduce polar co-ordinates with the polar axis in the direction of $\mathbf{K}$. The subsidiary angular variables of integration are denoted by θ' and ϕ' to distinguish them from θ (and ϕ) on which **K** depends and which are the physically significant angles through which the particle is actually scattered. Thus

$$\tilde{V}(K) = \int e^{i\mathbf{K}\cdot\mathbf{r}}\, V(r)\, d^3 r$$

$$= \int_0^{2\pi} \int_0^{\pi} \int_0^{\infty} e^{iKr\cos\theta'}\, V(r)\, r^2 \sin\theta'\, dr\, d\theta'\, d\phi'. \qquad (14.57)$$

The integration over ϕ' can be done immediately and the integration over θ' is also simple. We are then left with

$$\tilde{V}(K) = \frac{4\pi}{K} \int_0^{\infty} r \sin Kr\, V(r)\, dr. \qquad (14.58)$$

As an example we consider the screened "Coulomb" potential

$$V(r) = g\, e^{-\mu r}/r. \qquad (14.59)$$

The integration is again straightforward, leading to

$$\tilde{V}(K) = \frac{4\pi g}{K^2}\left(1 + \frac{\mu^2}{K^2}\right). \qquad (14.60)$$

The Coulomb potential for a particle of charge $Z_1 e$ scattering off a fixed charge $Z_2 e$ is obtained by taking

$$g = Z_1 Z_2 e_M^2, \qquad \mu = 0. \qquad (14.61)$$

Substituting into (14.51) gives the differential cross-section for Coulomb (or Rutherford) scattering to be

$$(\sigma(\theta) =)\, \frac{d\sigma}{d\Omega} = \left(\frac{2Z_1 Z_2 e_M^2 m}{\hbar^2 K^2}\right)^2. \qquad (14.62)$$

Squaring Eqn. (14.50) gives

$$K^2 \hbar^2 = 4p^2 \sin^2\theta/2, \qquad (14.63)$$

so

$$\frac{d\sigma}{d\Omega} = \left(\frac{Z_1 Z_2 e_M^2 m}{2p^2 \sin^2\theta/2}\right)^2, \qquad (14.64)$$

which for small θ is closely related to the approximate expression obtained from classical theory in Eqn. (10.16). It is evident from (14.60) that the fact that the quantum cross-section does not depend on $\hbar$ and is identical with the exact classical result is a peculiarity of a simple $1/r$ potential.

We remark finally that we have considered the scattering of a single particle in a fixed potential. The more usual physical problem is the scattering of two particles on each other under the action of a mutual interaction potential depending on the distance between them. As explained in § 10.5 the scattering in the Centre of Mass frame is then given by the identical formalism, provided the mass of the scattered particle is replaced by the reduced mass of the two particle system.

The total cross section is obtained by integrating over the solid angle.

§ 14.3 Radiative Transitions

The Golden Rule can also be used to calculate atomic transitions which are induced by electro-magnetic radiation. We consider absorption and stimulated emission resulting from a classical radiation field, since these problems can be handled without introducing the complication of photons. We deal explicitly with transitions between discrete states, paying particularly careful attention to the time dependence.

Consider a plane pulse of electromagnetic radiation of given duration and with a given frequency spectrum. This can be expressed in terms of a vector potential

$$\mathbf{A}(\mathbf{r},\,t) = \int_{-\infty}^{\infty} \mathbf{A}(\omega)\, e^{-i(\omega t - \mathbf{k}.\mathbf{r})}\, d\omega, \tag{14.65}$$

where $\mathbf{k}$ is a vector in the direction of propagation of the pulse with magnitude

$$k = \omega/c. \tag{14.66}$$

Since

$$\operatorname{div} \mathbf{A} = 0, \tag{14.67}$$

it follows that

$$\mathbf{A}.\mathbf{k} = 0 \tag{14.68}$$

and the reality of $\mathbf{A}(r\,.\,t)$ implies that

$$A(-\omega) = A^*(\omega). \tag{14.69}$$

The electric field is

$$\mathcal{E} = -\frac{\partial \mathbf{A}}{\partial t} \tag{14.70}$$

so

$$|\mathcal{E}| = \int_{-\infty}^{\infty} \omega A(\omega) e^{-i(\omega t - \mathbf{k}.\mathbf{r})}\, d\omega \tag{14.71}$$

and the magnetic induction is

$$\mathbf{B} = \text{curl } \mathbf{A}, \tag{14.72}$$

giving (using (14.66))

$$|\mathbf{B}| = \int_{-\infty}^{\infty} \frac{\omega}{c} A(\omega) e^{-i(\omega t - \mathbf{k} \cdot \mathbf{r})} d\omega. \tag{14.73}$$

The energy flux (energy per unit time per unit area perpendicular to the direction of propagation) is determined by the Poynting vector. This is parallel to k and of magnitude

$$|\mathbf{N}| = \left| \frac{1}{\mu_0} \mathscr{E} \wedge \mathbf{B} \right|$$

$$= \frac{1}{c\mu_0} \int\int \omega\omega' A(\omega) A(\omega') e^{-i[(\omega+\omega')t - (\mathbf{k}+\mathbf{k}')r]} d\omega \, d\omega'. \tag{14.74}$$

The total energy passing through unit area perpendicular to the direction of propagation from the whole duration of the pulse is (using (12.39))

$$\int_{-\infty}^{\infty} |N| \, dt$$

$$= \frac{2\pi}{c\mu_0} \int \int_{-\infty}^{\infty} \omega\omega' A(\omega) A(\omega') e^{i(\mathbf{k}+\mathbf{k}') \cdot \mathbf{r}} \delta(\omega+\omega') \, d\omega \, d\omega'$$

$$= \frac{4\pi}{c\mu_0} \int_{0}^{\infty} \omega^2 |A(\omega)|^2 \, d\omega. \tag{14.75}$$

The total energy from the full duration of the pulse per unit area in the frequency interval $\omega - \omega + d\omega$ is thus

$$I(\omega) \, d\omega = \frac{4\pi\omega^2}{c\mu_0} |A(\omega)|^2 \, d\omega = 4\pi\epsilon_0 c\omega^2 |A(\omega)|^2 \, d\omega, \tag{14.76}$$

where we have used the relation

$$c^2 \mu_0 \epsilon_0 = 1. \tag{14.77}$$

Notice that the intensity per unit area $I(\omega)$ has dimensions

$$[I(\omega)] \sim \frac{\text{energy} \times \text{time}}{\text{area}}. \tag{14.78}$$

We now calculate the probability that the pulse induces a transition in a hydrogen atom (with nucleus at rest at the origin) from an initial state

$$|i\rangle \equiv |n, l, m\rangle$$

to a final state

$$|f\rangle = |n', l', m'\rangle.$$

The process is emission or absorption according to whether E_i is of higher or lower excitation than E_f. The interaction energy between the radiation field and the atomic electron is

$$\hat{V} = \mathbf{A}(r, t) \cdot \hat{\mathbf{j}}$$

$$= \frac{e}{m_e} \mathbf{A}(r \cdot t) \cdot \hat{\mathbf{p}} \qquad (14.79)$$

where $\hat{\mathbf{j}}$ is the electron current density and $\hat{\mathbf{p}}$ is the momentum operator of the electron. Since $\hat{V}$ is explicitly time dependent, we cannot immediately use the Golden Rule, but must go back to (14.26). Substituting (14.79) for $\hat{V}$ and (14.65) for $\mathbf{A}$,

$$a_{fi}(T/2) = (i\hbar)^{-1} \left\langle f \left| \int \mathbf{A}(\omega) \frac{e}{m_e} \cdot \hat{\mathbf{p}} \, e^{i\mathbf{k} \cdot \mathbf{r}} \right| i \right\rangle$$

$$\times \int_{T/2}^{T/2} e^{i(E_f - E_i - \hbar\omega)t/\hbar} \, dt \, d\omega. \qquad (14.80)$$

Thus

$$a_{fi}(\infty) = -2\pi i \left\langle f \left| \int \mathbf{A}(\omega) \cdot \mathbf{p} \frac{e}{m_e} \, e^{i\mathbf{k} \cdot \mathbf{r}} \right| i \right\rangle$$

$$\times \delta(E_i - E_f - \hbar\omega) \, d\omega$$

$$= \frac{-2\pi i}{\hbar} \frac{e}{m_e} \mathbf{A}(\omega_{fi}) \langle f | \hat{p}_A \, e^{i\mathbf{k} \cdot \mathbf{r}} | i \rangle \qquad (14.81)$$

where in the final expression

$$\hbar\omega_{fi} = E_i - E_f. \qquad (14.82)$$

This is the Bohr rule (1.21).

The magnitude of k in the exponent of (14.81) is, by (1.27),

$$k = \frac{w_{fi}}{c} = \left(\frac{e_M^2}{\hbar c} \right) \frac{1}{2a_0} \left| \frac{1}{n^2} - \frac{1}{n'^2} \right| \qquad (14.83)$$

so

$$ka_0 \simeq \left(\frac{e_M^2}{\hbar c} \right) = \frac{1}{137}. \qquad (14.84)$$

This establishes the important result that the wave lengths of radiation emitted or absorbed in atomic transitions are two orders of magnitude larger than the size of the atom. The matrix element, more explicitly, is

$$\langle f | \hat{p}_A \, e^{i\mathbf{k} \cdot \mathbf{r}} | i \rangle = \int u_{n'l'm'}^*(\mathbf{r}) \, (-i\hbar \nabla_A) \, e^{i\mathbf{k} \cdot \mathbf{r}} u_{nlm}(\mathbf{r}) \, d^3 r. \qquad (14.85)$$

Since the eigen-functions $u(\mathbf{r})$ are essentially zero for r much greater than the Bohr radius, the relation (14.84) shows that the exponential in (14.81) may be approximated by unity. This is known as the dipole approximation.

Using the dipole approximation in (14.81), the probability that the pulse induces the transition is

$$
\begin{aligned}
w &= |a_{fi}(\infty)|^2 \\
&= \frac{4\pi^2}{\hbar^2} \frac{e^2}{m_e^2} |\mathbf{A}(\omega_{fi})|^2 \, |\langle f|\hat{p}_A|i\rangle|^2 \\
&= 4\pi^2 \left(\frac{e_M^2}{\hbar c}\right) \frac{I(\omega)}{m_e^2 \, \omega_{fi}^2 \hbar} |\langle f|\hat{p}_A|i\rangle|^2.
\end{aligned}
\tag{14.86}
$$

To obtain the final expression we have used (14.76) and the definition of e_M^2.

In principle this is the answer, but one can obtain a much deeper understanding from the following simple argument.

From dimensional considerations it is reasonable to approximate the momentum by†

$$
\langle f|\hat{p}_A|i\rangle \simeq m_e \omega_{fi} a_0.
\tag{14.87}
$$

Substituting this into (14.86),

$$
w \simeq 4\pi \left(\frac{e_M^2}{\hbar c}\right) \left(\frac{I(\omega_{fi}) \, \pi a_0^2}{\hbar}\right).
\tag{14.88}
$$

† This somewhat brash argument can be substantiated as follows.
By the correspondence principle

$$
\hat{\mathbf{p}}_A = m_e \frac{d\hat{\mathbf{r}}_A}{dt}
$$

where $\mathbf{r}_A$ is the component of the position operator of the electron in the direction of the vector potential. By (13.37)

$$
i\hbar \frac{d\hat{\mathbf{r}}_A}{dt} = [\hat{\mathbf{r}}_A, \hat{H}]
$$

so

$$
\begin{aligned}
\langle f|\hat{\mathbf{p}}_A|i\rangle &= (i\hbar)^{-1} m_e \langle f|\hat{\mathbf{r}}_A \hat{H} - \hat{H} r_A|i\rangle \\
&= i m_e \frac{(E_f - E_i)}{\hbar} \langle f|\hat{\mathbf{r}}_A|i\rangle \\
&= i m_e \omega_{fi} \langle f|\hat{\mathbf{r}}_A|i\rangle.
\end{aligned}
$$

It is now extremely plausible to approximate

$$
\langle f|\hat{\mathbf{r}}_A|i\rangle \simeq a_0
$$

so that

$$
|\langle f|\hat{\mathbf{p}}_A|i\rangle| \simeq m_e \omega_{fi} a_0.
$$

Written in this form the expression has a very clear physical interpretation. Since we are calculating a probability the result must be a pure number. The first factor in brackets is the fine structure constant (see (8.60)) which expresses in a dimensionless way the strength of the electromagnetic interaction. We have thus established a result anticipated in the general discussion of § 8.6. In the final factor πa_0^2 is clearly the cross-sectional area of the atom as viewed from the beam. Since $I(\omega)$ is an intensity per unit area, the numerator is, therefore, the energy intensity of the entire pulse at the appropriate Bohr frequency which strikes the atom. It is evident from (14.78) that this numerator has the same dimensions as h, so the whole factor is dimensionless—as it must be.

The "cross section" for the transition which is independent of the pulse is defined by multiplying (14.86) by $\hbar\omega$ to give the probable energy loss and then dividing by the intesnity factor. Thus (using the expression from the footnote)

$$\sigma_{fi} \equiv \frac{\hbar\omega_{fi}w}{I(\omega_{fi})} = 4\pi\omega_{fi}\left(\frac{e_M^2}{\hbar c}\right) . \pi|\langle f|\hat{r}_A|i\rangle|^2. \tag{14.89}$$

§ 14.4 Beta Decay

We now apply the Golden Rule to the nuclear problem of β-decay to establish in more detail the statement made in § 11.4 that the observed mean lives of the unstable sub-nuclear particles have to be interpreted in terms of a separate Weak nuclear interaction. We consider the specific process of neutron decay

$$n \to p + e^- + \bar{\nu}.$$

This is historically the first effect of the Weak interaction to be discovered and it is evident from Table 11.1 that it is doubly anomalous in that the observed mean life of eleven minutes differs by twenty-seven orders of magnitude from the typical nuclear time. The explanation of this very large factor brings out yet another important point.

We consider the decay of a free neutron at rest and neglect spin. Since the neutron and proton are so much heavier than leptons (electron and neutrino) it is reasonable to neglect recoil and suppose that the proton is also at rest. The nucleons may be described by highly localized normalized wave functions

$$u_n(r), \qquad u_p(r), \tag{14.90}$$

which vanish for distances appreciably greater than the nucleon Compton radius; that is for

$$r > \hbar/m_p c. \tag{14.91}$$

The electron and neutrino are emitted with high momenta, p_v and p_e respectively, and their wave functions are the corresponding de Broglie waves which we shall again normalize in a large volume L^3 (see (14.40)). We take the interaction which causes the transition to be specified simply by a parameter g_β which determines its strength. Thus the matrix element which appears in the Golden Rule is

$$|\langle i | V | f \rangle|^2 = \left| \frac{g_\beta}{L^3} \int u_n^*(r) \, u_p(r) \, e^{ip_e r/\hbar} e^{ip_v r/\hbar} d^3 r \right|^2. \tag{14.92}$$

Since

$$p_e \ll m_p c, \tag{14.93}$$

it follows that

$$p_e r/\hbar \ll 1 \tag{14.94}$$

for all values of r for which $u_p(r)$ is non-zero. The same is also true for p_v, so both the exponentials in the integral can be replaced by unity. Since we may expect $u_n(r)$ and $u_p(r)$ to be similar functions, the integral now approximates to the normalization integral which can also be taken to be unity. It is convenient to define the strength of the Weak interaction in terms of a constant f^2, which has the same dimensions as e^2. To do this we must divide g_β by a squared length which it is reasonable to take as the proton Compton radius. Thus

$$g_\beta = f^2 (\hbar/m_p c)^2, \tag{14.95}$$

and in this approximation

$$|\langle f | V | i \rangle|^2 = \frac{f^4}{L^6} \left(\frac{\hbar}{m_p c} \right)^4. \tag{14.96}$$

The number of states in the phase space available to the electron and to the neutrino is exactly the same as for the single particle in the scattering problem. Since the proton is in a unique state, we have, by (14.41), that

$$\rho(E_f) = \delta(E_i - E_f) \frac{L^3 p_e^2 dp_e d\Omega_e}{(2\pi\hbar)^3} \frac{L^3 p_v^2 dp_v d\Omega_v}{(2\pi\hbar)^3}. \tag{14.97}$$

It is convenient to write

$$dp_e \, dp_v = \left(\frac{\partial p_v}{\partial E_f} \right) dp_e \, dE_f \tag{14.98}$$

and integrate over E_f. This removes the δ-function from the expression but enforces energy conservation. Thus

$$E_i = m_n c^2$$
$$= E_f = m_p c^2 + \sqrt{p_e^2 c^2 + m_e^2 c^4} + p_\nu c, \qquad (14.99)$$

and

$$\frac{\partial E_f}{\partial p_\nu} = c, \qquad (14.100)$$

giving

$$\rho(E_f) = \left(\frac{L}{2\pi\hbar}\right)^6 \frac{1}{c} p_e^2 p_\nu^2 \, dp_e \, d\Omega_e \, d\Omega_\nu. \qquad (14.101)$$

We can now substitute (14.96) and (14.101) into the Golden Rule (14.35) to give the differential rate for β-decay

$$\frac{d^3 w}{dp_e \, d\Omega_e \, d\Omega_\nu} = \frac{2\pi}{\hbar} \left(\frac{f\hbar}{m_p c}\right)^4 \left(\frac{1}{2\pi\hbar}\right)^6 \frac{p_e^2 p_\nu^2}{c}. \qquad (14.102)$$

This pertains to the most complete observation which can in principle be made, in which the directions of both leptons and the energy of the electron are measured, the energy of the neutrino then being determined by energy conservation. The angular measurements could give information about the angular dependence of the interaction. Since we have already assumed isotropy nothing is lost by integrating over both solid angles, each of which gives a factor of 4π. Thus, after some re-arranging,

$$\frac{dw}{dp_e} = \frac{1}{2\pi^3} \left(\frac{f^2}{\hbar c}\right)^2 \left(\frac{m_p c^2}{\hbar}\right) \frac{p_e^2 p_\nu^2}{(m_p c)^5}, \qquad (14.103)$$

which determines the momentum spectrum of the electron. The maximum value for the electron momentum, p_{max}, is found by putting the neutrino momentum zero in (14.99):

$$(m_n - m_p)c = \sqrt{p_2^{max} + m_e^2 c^2}. \qquad (14.104)$$

Then, eliminating p_ν from (14.103),

$$\frac{dw}{dp_e} \sim p_e^2 (\sqrt{p_{max}^2 + m_e^2 c^2} - \sqrt{p_e^2 + m_e^2 c^2})^2 \qquad (14.105)$$

which shows that the electron may be emitted with any momentum between zero and the maximum, the most probably value lying near the middle of the allowed range. In marked contrast, if there were no neutrino in the final state, the momentum of the electron would be uniquely determined by energy conservation. The observation of the electron spectrum in β-decay was, historically, the first evidence for the existence of the elusive neutrino.

To calculate the total rate the partial rate (14.103) must be integrated over p_e up to the maximum value given by (14.104),

$$p_{\max} \simeq \tfrac{3}{2} m_e c. \tag{14.106}$$

Since, by (14.105), $m_e c$ is also the typical momenta in the integrand we may approximate the integral, on dimensional grounds, by

$$\int_0^{p_{\max}} p_e^2 p_\nu^2 dp_e \simeq (m_e c)^5. \tag{14.107}$$

Combining this with (14.103) we obtain for the rate

$$w \simeq \frac{1}{2\pi^3} \left(\frac{m_p c^2}{\hbar}\right) \left(\frac{f^2}{\hbar c}\right)^2 \left(\frac{m_e}{m_p}\right)^5. \tag{14.108}$$

Each of the three factors in brackets has a direct and simple physical interpretation. The first is the inverse of the typical nuclear time (see (11.59)), which has to be there for dimensional reasons, since w is a probability per unit time. The second factor is a dimensionless constant which fixes the strength of the β-decay (Weak nuclear) interaction in a manner exactly analogous to the fine structure constant in electromagnetism. The final factor is also clearly dimensionless and is a measure on an appropriate nuclear scale of the amount of phase space available to the system in the final state. As remarked earlier we have to account for the enormous discrepancy between the observed rate which is

$$w = 10^{-3} \sec^{-1} \tag{14.109}$$

and the inverse nuclear time,

$$\frac{m_p c^2}{\hbar} \simeq 1\cdot 4 \times 10^{24} \sec^{-1}. \tag{14.110}$$

This is partially explained by the phase space factor

$$\left(\frac{m_e}{m_p}\right)^5 \simeq 10^{-16} \tag{14.111}$$

which expresses the fact that the neutron is only just sufficiently heavier than the proton for the decay to be energetically possible. Substituting these values into (14.108) shows that

$$(f^2/\hbar c^2) \simeq 10^{-10}, \tag{14.112}$$

which is the strength of the Weak interaction stated in Table 11.2. (The fact that it appears squared is purely conventional.) In the other decays listed in Table 11.1 the amount of energy available is much greater, the phase space factor is much closer to unity, so that the life times which are very long on a nuclear scale (but not nearly as long as that of the neutron) are controlled mainly by the weakness of the Weak interaction (see problem **14.3**).

We remark that the same analysis applies to the beta-decay of a neutron (or proton) inside a nucleus provided the rest energies of the nucleons which appear in the energy conservation condition (14.99) are replaced by the energies of the parent and daughter nuclei in the decay. This changes the definition of p_{max} (14.104) and may very substantially change the contribution from the phase space factor. Finally, it should be pointed out that since we have neglected spin we have not been able to discuss further the parity non-conservation effects described at the end of § 13.3.

§ 14.5 Summary

Using the Schrödinger equation of motion we have derived the Golden Rule—an approximate expression for the probability rate for transitions of a system from one unperturbed state to another.

This formula has first been used to calculate scattering cross sections in a given potential; in particular the cross section for Rutherford scattering. It has then been applied in a semi-classical treatment to radiative transitions of atoms, showing that these are proportional to the fine structure constant $(e_M^2/\hbar c)$ multiplied by the effective area of the atom.

Finally we have used the Golden Rule to establish quantitatively the physical significance of the beta-decay of a free neutron—the basic beta-decay process. Most important, we have established the weakness of the Weak interaction $(f^2/\hbar c)$, but we have also shown that decay rates can be very significantly affected by the amount of phase space available to the decaying system in the final state.

PROBLEMS XIV

14.1. By considering the angular part of the integral in (14.85) show that in the dipole approximation the transition probability for emission or absorption for a hydrogen atom vanishes unless $l - l' = \pm 1$ and $m = m'$ or $m - m' = \pm 1$.

14.2. The interaction between a free neutron and a heavy nucleus may be described in terms of a "square well" potential

$$V(r) = -V_0 \qquad r \leqslant a$$
$$V(r) = 0 \qquad r > a.$$

Show that at high energy the differential cross section is proportional to $(j_1(Ka)/Ka)^2$, where $\mathbf{K}$ is the momentum transfer and the spherical Bessel function is

$$j_1(Ka) = [\sin (Ka) - Ka \cos (Ka)]/(Ka)^2.$$

Show that the differential cross section has a strong forward peak with subsidiary maximum at

$$Ka \simeq 6.$$

14.3. According to Table 11.1 a Λ-hyperon decays into a proton and π-meson (pion). Using exactly the same approximation as for β-decay (i.e. neglecting the recoil of the heavier particles) show that the energy available to the pion is approximately $(5/4)m_\pi c^2$ and that its momentum is about $(3/4)m_\pi c$. For dimensional reasons the decay interaction must be

$$V^2 \sim f_\Lambda^4 \left(\frac{\hbar}{m_p c} \right).$$

Apply the Golden Rule to show that in this approximation the total decay rate (inverse of the mean life) is

$$\omega = \frac{15}{16\pi} \left(\frac{f_\Lambda^2}{\hbar c} \right)^2 \left(\frac{m_p c^2}{\hbar} \right) \left(\frac{m_\pi}{m_p} \right)^2.$$

By comparison with (14.108) note that the change in the final phase space factor accounts for the enormous difference between the Λ and neutron mean lives assuming roughly equal strengths for the Weak decay interaction in the two cases.

CHAPTER 15

UNITARY SYMMETRY AND
SUB-NUCLEAR PHYSICS

§ 15.1 Strong Interactions, Electric Charge, Baryon Charge and Hypercharge

We are now in a position to continue the discussion of the properties of sub-nuclear particles, which is started in § 11.4. It is shown there that a large number of sub-nuclear particles are found in proton–proton collisions. These are the mesons, nucleons (neutron and proton) and hyperons listed in Table 11.1. These particles are produced, and interact with each other, through the *strong* nuclear interaction, and are known as the Strongly Interacting Particles, or *hadrons*. Most of them are unstable and subsequently disintegrate, or decay, through the *weak* nuclear interactions discussed in § 14.4. We are primarily concerned here with the *strong* interaction.

We have mentioned proton–proton collisions. In modern proton accelerators bunches of about 10^{12} protons can be accelerated every few seconds to an energy of 500,000 MeV (=500 GeV). At these energies and intensities the production of secondary particles $\pi^{\pm}$, $\kappa^{\pm}$ and anti-protons, $\bar{p}$, is so copious that these particles can be separated electromagnetically into beams, and directed into hydrogen bubble chambers. The charged particles produced in collisions with the protons in the hydrogen make tracks which can be photographed. The bubble chambers are placed in strong magnetic fields which produce a curvature of these tracks and enable the experimenter to analyse the collisions in terms of the energy, momentum, and mass of the particles involved. In this way a great variety of sub-nuclear collisions have been studied under laboratory conditions. From these, a number of regularities have been found in the form of conservation laws.

The first of these is rather obvious. We are considering essentially nuclear interactions, so that to a good approximation electric and magnetic forces can be neglected. However, the particles in the initial and final states each have electric charge, which has an integer value —usually plus or minus one, or zero—in units of the charge on the electron. In any nuclear collision, or decay process, the net charge of the particles before and after is the same. Thus, for example, in proton–proton collisions in which pions are produced, we find

$$pp \to pn\pi^+,$$

but not

$$pp \not\to pp\pi^-$$

If we introduce a charge operator $\hat{Q}$ representing the observation of the net charge of the system, it is conserved. Thus, by (13.47),

$$[\hat{H}_s, \hat{Q}] = 0, \tag{15.1}$$

where $\hat{H}_s$ is the Hamiltonian which describes the strong nuclear interaction of hadrons. We express the laws of conservation of electric charge as

$$\Delta Q = 0. \tag{15.2}$$

The next law is somewhat similar. It follows from the fact that the proton is stable, and does not decay into any combination of the many lighter particles which are available (see Table 11.1). The simplest way to ensure this is to attribute to the proton and to all (half-integer spin) particles heavier than the proton, a different sort of charge called baryon charge, $\hat{B}$ which also takes integer values. Thus in Table 11.1 p, n, Λ, Σ and Ξ all have $\langle B \rangle = 1$. The mesons, κ, η and π have $\langle B \rangle = 0$. For anti-particles the baryon-charge, like electric charge, is equal in magnitude but opposite in sign to that of the particles. We postulate that the net baryon charge is also conserved in any nuclear interaction.

$$\Delta B = 0. \tag{15.3}$$

Thus we may have

$$pp \to p\Sigma^+ \kappa^0. \tag{15.4}$$

$$\to pp\eta, \tag{15.5}$$

$$\pi^- p \to p\Lambda\bar{\Lambda}\pi^-, \tag{15.6}$$

but not

$$\pi^+ p \not\to \pi^+ \kappa^+ \bar{\kappa}^0.$$

The reader may check for himself that this rule effectively prevents any chain of interactions, which would enable protons to disintegrate into lighter particles. These two conservation laws of electric and baryon charge apply to all interactions—strong, electromagnetic and weak.

It has been found that there are also more subtle regularities in the purely strong production collisions, which are not accounted for by these two rules. They can be systematized by attributing to each hadron yet another type of charge, the *hypercharge*, $\hat{Y}$. The value of Y for the different hadrons is given in Table 15.1, with again the rule

TABLE 15.1. *Sub-nuclear charges*

The table gives the electric charge Q, the baryon charge B, and the hypercharge Y of the longest known sub-nuclear particles. The isotopic spin component is

$$I_3 = Q - \frac{Y}{2}.$$

	B	Q	Y	$I_3 = Q - \dfrac{Y}{2}$
$\varXi^-$	1	-1	-1	$-\frac{1}{2}$
$\varXi^0$	1	0	-1	$+\frac{1}{2}$
$\varSigma^+$	1	$+1$	0	$+1$
$\varSigma^0$	1	0	0	0
$\varSigma^-$	1	-1	0	-1
$\varLambda$	1	0	0	0
n	1	0	$+1$	$-\frac{1}{2}$
p	1	$+1$	$+1$	$+\frac{1}{2}$
κ^-	0	-1	-1	$-\frac{1}{2}$
κ^0	0	0	-1	$+\frac{1}{2}$
π^+	0	$+1$	0	$+1$
π^0	0	0	0	0
π^-	0	-1	0	-1
η	0	0	0	0
κ^0	0	0	$+1$	$-\frac{1}{2}$
κ^+	0	$+1$	$+1$	$+\frac{1}{2}$

that the hypercharge of any anti-particle is equal in magnitude but opposite in sign to that of the corresponding particle. We postulate further that the net hypercharge is conserved in any nuclear collision to the extent that electromagnetic and weak interactions can be neglected:

$$\Delta Y = 0. \tag{15.7}$$

Thus we may have

$$p\pi^+ \to \varSigma^+ \kappa^+ \pi^0, \tag{15.8}$$

$$p\bar{p} \to \varLambda \bar{\varLambda}, \tag{15.9}$$

$$p\pi^- \to \varXi^- \kappa^+ \kappa^0; \tag{15.10}$$

but not, for example

$$\kappa^- p \not\to \pi^- p.$$

Thus to every sub-nuclear particle, in addition to its mass and spin, we attribute three charges—the electric charge, the baryon charge and the hypercharge—and the net total of each of these is found to be conserved in any strong interaction.

These conservation laws can be related to invariance with respect to unitary transformations in the following manner (see § 13.5). The state functions describing the initial and final states of the collision processes considered above are considerably more complicated than those encountered in previous chapters. In addition to giving information on the spatial and spin configurations of the particles, they must also specify their nature; that is whether they are protons $|p\rangle$ or neutrons $|n\rangle$ and so forth. Similarly, the interaction Hamiltonian must be expressible in terms of operators which annihilate and create these particles, so that as time passes the nature of the particles in the state can change. (This is a generalization of the expressions (12.71) and (12.72) for the energy of a harmonic oscillator in terms of annihilation and creation operators of units of energy.) We are here concerned only with these latter factors in the state vectors, which must be eigenstates of $\hat{Q}$, $\hat{B}$ and $\hat{Y}$. If the initial state in a collision is $|i\rangle$, then for example,

$$\hat{Q}|i\rangle = Q_i|i\rangle \tag{15.11}$$

where Q_i is the net total charge of the initial state.

The corresponding unitary transformation

$$\hat{U}_Q = \exp[i\hat{Q}\epsilon] \tag{15.12}$$

where ϵ is a real parameter. If $\hat{H}_s$ is the Hamiltonian for the strong interaction, the conservation of charge is expressed as the invariance of H_s with respect to (15.12) (see (13.68))

$$\langle f|H_s|i\rangle = \langle f|U_Q^+ H_s U_Q|i\rangle$$
$$= \langle f|H_s|i\rangle \exp[i(Q_i - Q_f)\,\epsilon]. \tag{15.13}$$

This implies that

$$\langle f|H_s|i\rangle = 0, \tag{15.14}$$

unless

$$Q_f = Q_i. \tag{15.15}$$

This says that the Hamiltonian only couples a final state $|f\rangle$ to an initial state $|i\rangle$ if they have the same total charge.

Note that if a state contains several charged particles, the effect of the charge operator is just to extract the net charge of the state:

$$U_Q |pn\pi^+\rangle = \exp[i\hat{Q}\epsilon]|pn\pi^+\rangle = \exp[i(Q_P + Q_\pi)\epsilon]|pn\pi^+\rangle. \quad (15.16)$$

where

$$Q_P = Q_\pi = +1$$

are the charges of the proton and π^+ meson, respectively. These transformations are one parameter unitary transformations and form what is known as the group U(1). We can introduce similarly unitary transformations $\hat{U}_B$ and $\hat{U}_Y$ associated with the conservation of baryon charge and hypercharge. At this stage these invariance properties of the Hamiltonian are not very useful, but they point the way to an extremely fruitful generalization.

§ 15.2 Isotopic Spin and SU(2)

A striking feature of the hadrons, which is not incorporated in the theory so far presented, is the way in which they appear in multiplets of different charge but very nearly equal mass (see Table 11.1). Thus there are two nucleons, three Σ's, two Ξ's. It is reasonable to suppose that the mass differences within a multiplet are electromagnetic effects, and that in the limit of purely strong interactions all members of a multiplet would have exactly the same mass. We suppose further that, in a sense to be made precise below, the strong interactions are invariant for interchange of particles within a mass multiplet.

To be specific, let us suppose that the proton and neutron are two possible states of a nucleon, and analogous to the two possible spin states of a proton. We define a two component nucleon spinor N_a ($a = 1, 2$),

$$N_a = \begin{pmatrix} p \\ n \end{pmatrix}. \quad (15.17)$$

Such factors appear in the state vectors to specify the nature of the particles. Suppose that the strong interaction is invariant for (2×2) unitary transformations of these spinors. The most general such transformation is

$$N_a \to \sum \hat{U}_a{}^b N_b,$$

where

$$\hat{U}_a{}^b = \exp\left[\frac{i}{2}(\epsilon^{(1)}\hat{\tau}_1 + \epsilon^{(2)}\hat{\tau}_2 + \epsilon^{(3)}\hat{\tau}_3)\right]_a^b, \quad (15.18)$$

and $\epsilon^{(i)}(i=1,2,3)$ are arbitrary real parameters. The three Hermitian matrices $(\hat{\tau}_i)_a{}^b$ are numerically identical with the Pauli spin matrices (8.32),

$$\hat{\tau}_1 = \begin{pmatrix} 0 & 1 \\ 1 & 0 \end{pmatrix}, \quad \hat{\tau}_2 = \begin{pmatrix} 0 & -i \\ i & 0 \end{pmatrix}, \quad \hat{\tau}_3 = \begin{pmatrix} 1 & 0 \\ 0 & -1 \end{pmatrix}. \quad (15.19)$$

We have excluded from these considerations transformations of the form

$$\hat{U} = \exp[i\epsilon\hat{1}], \quad \hat{1} = \begin{pmatrix} 1 & 0 \\ 0 & 1 \end{pmatrix}, \quad (15.20)$$

which are just simultaneous phase transformations of p and n. These have already been introduced in connection with baryon charge conservation.) Since we are dealing with (2×2) unitary transformations, we are considering the group of transformations SU(2). Because of the analogy with spin, the transformations are known as *isotopic spin* transformations.

Of the three $\hat{\tau}$ matrices,

$$\tfrac{1}{2}\hat{\tau}_3 \equiv \hat{I}_3 \quad (15.21)$$

is diagonal. By comparing (15.18) with (13.57) for transformations in which only $\epsilon^{(3)}$ is non-zero, we see that $\hat{I}_3$ plays the role of $\hat{F}$, and is a conserved observable similar to $\hat{B}$, $\hat{Q}$ and $\hat{Y}$. By direct substitution of the explicit matrices, and state vectors

$$\exp[i\epsilon^{(3)}\,\hat{I}_3]|p\rangle = \exp[i\epsilon^{(3)}(+\tfrac{1}{2})]|p\rangle \quad (15.22)$$

and

$$\exp[i\epsilon^{(3)}\,\hat{I}_3]|n\rangle = \exp[i\epsilon^{(3)}(-\tfrac{1}{2})]|n\rangle \quad (15.23)$$

so that p and n have I_3 "charge" $+\tfrac{1}{2}$ and $-\tfrac{1}{2}$ respectively.

We must now construct other multiplets of particles which transform among themselves under these isotopic spin transformations. Directly related to the nucleon spinor is the anti-nucleon spinor

$$\bar{N}^a = (\bar{p}, \bar{n}), \quad (15.24)$$

which transforms

$$\bar{N}^a \to \sum_b \bar{N}^b \, \hat{U}^+{}_b{}^a, \quad (15.25)$$

so the I_3 values of $\bar{p}$ and $\bar{n}$ are $-\tfrac{1}{2}$ and $+\tfrac{1}{2}$ respectively. Other multiplets may be obtained by combining nucleons and anti-nucleons. Thus we have the scalar combination

$$\eta = \sum_a \frac{1}{\sqrt{2}} N_a \bar{N}^a = \frac{p\bar{p}+n\bar{n}}{\sqrt{2}}, \quad (15.26)$$

with $Q=0$, $B=0$, $Y=0$. We also have

$$\pi_a{}^b \equiv [N_a \bar{N}^b - \tfrac{1}{2}\delta_a{}^b (\sum_c N_c \bar{N}^c)]$$

$$= \begin{pmatrix} \dfrac{p\bar{p}-n\bar{n}}{2}, & p\bar{n} \\ n\bar{p}, & -\dfrac{p\bar{p}-n\bar{n}}{2} \end{pmatrix}$$

$$\equiv \begin{pmatrix} \pi^0/\sqrt{2} & \pi^+ \\ \pi^- & -\pi^0/\sqrt{2} \end{pmatrix} \tag{15.27}$$

The Q, B, Y and I_3 values of the different nucleon–anti-nucleon combinations are determined by simply adding the appropriate values of the constituent particles. We may then identify the three different combinations, which appear with the three states of the π-meson, π^+, π^0, π^-, with I_3 values $+1$, 0, -1, respectively. This is the significance of the final expression in (15.27). These multiplets may be thought of quite physically as bound states made up of nucleons and anti-nucleons, interacting through forces, which are invariant with respect to isotopic spin transformations.

These three forms of multiplet—scalar (singlet), spinor (doublet) and vector (triplet)—are sufficient to account for all the sub-nuclear particles given in Table 11.1, and they are shown graphically, with their Y values in Fig. 15.1. Displayed in this way they form regular patterns which are discussed in the next section.

Before proceeding to this discussion we consider an alternative geometrical construction of the SU(2) (isotopic) multiplets. Specify a multiplet graphically by the I_3 values of the components. The simplest is the single component scalar, $I_3=0$. Fig. 15.2a, and the next is the spinor $I_3 = \pm\tfrac{1}{2}$, Fig. 15.2b. Since I_3 is additive, when two spinors are combined we can either add or subtract $\tfrac{1}{2}$ from one of them to each of the components of the other, to obtain the I_3 values of the combination. This is represented in Fig. 15.3, in which the diagram for the basic spinor is superposed on the two points of Fig. 15.2b. Removing one of the two resulting points at the origin, since it reproduces Fig. 15.2a (the scalar of (15.26)), we are left with the triplet constructed algebraically in (15.27), ($I_3 = \pm 1, 0$). If a further doublet is combined in this way with the triplet, the reader may easily see that this reproduces the original doublet, and a new

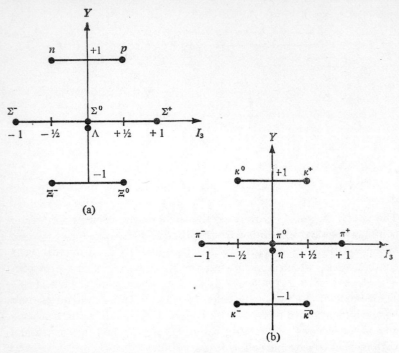

FIG. 15.1. Diagrams showing the isotopic, SU(2) particle sub-multiplets (joined by heavy lines) with their I_3 and Y values. All the particles in (a) have $B = 1$ and spin $\frac{1}{2}\hbar$, while those in (b) have $B = 0$ and spin 0. Each comprises an eight-fold multiplet of SU(3). A similar multiplet of particles of $B = 0$, spin $\hbar$ were found in 1961. These are obtained by the replacement $\pi \rightarrow \rho$ (765 MeV/c²) $\eta \rightarrow \phi$ (1020 MeV/c²), $\kappa \rightarrow \kappa^*$ (890 MeV/c²).

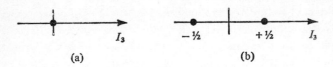

FIG. 15.2. Graphical representation of (a) the scalar singlet and (b) the basic spinor doublet of SU(2).

multiplet of four states with $I_3 = \pm \frac{3}{2}$, $\pm \frac{1}{2}$. This procedure can clearly be extended and is generalized in the next section to the construction of multiplets corresponding to more complicated invariance properties.

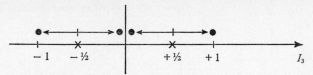

FIG. 15.3. Geometrical construction of the multiplets obtained by the combination of two SU(2) doublets. The diagram for one doublet (Fig. 15.2b) is superposed on each member of the other (marked ✕) to give a singlet ($I_3 = 0$) and a triplet ($I_3 = \pm 1, 0$).

§ 15.3 The Eight-fold Way and SU(3)

When the SU(2) multiplets of particles of the same spin and baryon charge are plotted as in Fig. 15.1 they make regular hexagonal patterns of eight particles. We wish to find an explanation for this.

The first point to notice from Table 15.1 is that

$$I_3 = Q - \frac{Y}{2}. \tag{15.28}$$

We introduced invariance with respect to SU(2) in § 15.2 to provide a basis for the sub-nuclear mass multiplets. This led to the conservation of I_3. If $\hat{Y}$ also is conserved, then by (15.28) the conservation of Q follows. From the point of view of the strong interactions the electric charge appears as a secondary quantity. To explain the regular patterns of Fig. 15.1 it is natural to extend these notions, and to try to incorporate $\hat{Y}$, too, into the invariance structure. To do this we must suppose that the strong interactions are at least approximately invariant under a wider group of transformations which gives rise to a second diagonal Hermitian matrix (conserved observable), which can be identified with $\hat{Y}$.

The simplest scheme which proves to work is that all sub-nuclear particles may be thought of as composed of a triplet of basic " quarks ",

$$q_a = \begin{pmatrix} p' \\ n' \\ \lambda' \end{pmatrix} \quad (a = 1, 2, 3), \tag{15.29}$$

and that the strong interactions are approximately invariant under unitary (3×3) transformations on this triplet, SU(3),

$$q_a \rightarrow \hat{U}_a{}^b q_b. \tag{15.30}$$

(The invariance can only be approximate since it leads to larger mass multiplets, combining together particles whose observed physical masses are quite widely separated.) The general transformation can be written

$$\hat{U} = \exp\left[i \sum_{j=1}^{8} \epsilon^{(j)} \hat{F}_j \right] \qquad (15.31)$$

where $\epsilon^{(j)}$ are arbitrary real parameters and the Hermitian matrices

$$(\hat{F}_j)_a{}^b \quad (a, b = 1, 2, 3) \qquad (15.32)$$

are given in Table 15.2. Of these, two are diagonal and give rise to additive conserved quantum numbers. From Table 15.2 it is evident that $\hat{F}_1$, $\hat{F}_2$ and $\hat{F}_3$ operate only on the first two components of q_a

TABLE 15.2. *The matrices of* SU(3)

The matrices which generate the transformations of SU(3). There are two diagonal matrices giving rise to additive conserved quantum numbers, which are identified as $\hat{F}_3 = I_3$, $\hat{F}_8 = \dfrac{\sqrt{3}}{2}\hat{Y}$.

$$\hat{F}_1 = \tfrac{1}{2}\begin{pmatrix} 0 & 1 & 0 \\ 1 & 0 & 0 \\ 0 & 0 & 0 \end{pmatrix}, \quad \hat{F}_2 = \tfrac{1}{2}\begin{pmatrix} 0 & -i & 0 \\ i & 0 & 0 \\ 0 & 0 & 0 \end{pmatrix}, \quad \hat{F}_3 = \tfrac{1}{2}\begin{pmatrix} 1 & 0 & 0 \\ 0 & -1 & 0 \\ 0 & 0 & 0 \end{pmatrix}$$

$$\hat{F}_4 = \tfrac{1}{2}\begin{pmatrix} 0 & 0 & 1 \\ 0 & 0 & 0 \\ 1 & 0 & 0 \end{pmatrix}, \quad \hat{F}_5 = \tfrac{1}{2}\begin{pmatrix} 0 & 0 & -i \\ 0 & 0 & 0 \\ i & 0 & 0 \end{pmatrix}, \quad \hat{F}_6 = \tfrac{1}{2}\begin{pmatrix} 0 & 0 & 0 \\ 0 & 0 & 1 \\ 0 & 1 & 0 \end{pmatrix}$$

$$\hat{F}_7 = \tfrac{1}{2}\begin{pmatrix} 0 & 0 & 0 \\ 0 & 0 & -i \\ 0 & i & 0 \end{pmatrix}, \quad \hat{F}_8 = \frac{1}{2\sqrt{3}}\begin{pmatrix} 1 & 0 & 0 \\ 0 & 1 & 0 \\ 0 & 0 & -2 \end{pmatrix}.$$

and reproduce the isotopic transformations of SU(2). The new group of transformations thus includes those of § 15.2 and in particular we establish the I_3 properties of the quarks listed in Table 15.3. The other diagonal matrix is $\hat{F}_8$. Our object is to relate this to $\hat{Y}$. The identification, which we justify *post hoc* by results, is

$$\hat{F}_8 = \frac{\sqrt{3}}{2}\hat{Y}. \qquad (15.33)$$

By considering the effect on q_a of transformations for which only $\epsilon^{(8)}$ is non-zero we can identify, as in (15.22) and (15.23), the Y values of p', n' and λ' given in Table 15.3. The relation (15.28) then determines the Q-values given in the Table. Finally we must take $B = \tfrac{1}{3}$, since

TABLE 15.3. *Properties of the quarks*

	I_3	Y	Q	B
p'	$+\frac{1}{2}$	$\frac{1}{3}$	$\frac{2}{3}$	$\frac{1}{3}$
n'	$-\frac{1}{2}$	$\frac{1}{3}$	$-\frac{1}{3}$	$\frac{1}{3}$
λ'	0	$-\frac{2}{3}$	$-\frac{1}{3}$	$\frac{1}{3}$

we wish to construct particles with integer hypercharge, and baryon charge equal to one. The anti-quark

$$\bar{q}^a = (\bar{p}', \bar{n}', \bar{\lambda}') \tag{15.34}$$

has all its quantum numbers, I_3, Y, Q, B, equal in magnitude, but opposite in sign to q_a. These two basic triplets are represented graphically in Fig. 15.4.

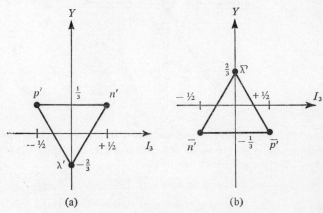

(a)   (b)

FIG. 15.4. A graphical representation of the basic SU(3) triplets; (a) the quark, **3**, and (b) the anti-quark, $\overline{\textbf{3}}$. If the scale is chosen the same for $F_3 = I_3$ and $F_8 = \dfrac{\sqrt{3}}{2} Y$, they are represented by equilateral triangles.

We can now apply the obvious generalization of the graphical technique of the previous section to construct other multiplets of particles which transform among themselves under the transformations of SU(3). Since I_3 and Y are additive "charge"-like quantum numbers, we construct the particles obtained by combining, for example q_a and $\bar{q}^a$ by superposing on each point of q_a (Fig. 15.4a) the triangle representing $\bar{q}^a$ (Fig. 15.4b). This is shown in Fig. 15.5. Of the three particles at the centre, one must be removed, since it represents the singlet scalar. The remaining points just reproduce the hexagonal eight-fold pattern of Fig. 15.1b. Since the baryon charge is $B = \frac{1}{3} - \frac{1}{3} = 0$, all the quantum numbers of the meson multiplet are satisfactorily accounted for. This combination of a triplet, **3**, and an

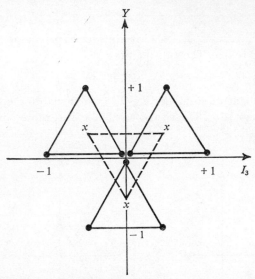

FIG. 15.5. The singlet and hexagonal octet multiplet pattern obtained by superposing the triangles of $\overline{\mathbf{3}}$ on the points of **3**, thus establishing graphically the multiplet formula

$$\mathbf{3} \otimes \overline{\mathbf{3}} = \mathbf{8} \oplus \mathbf{1}.$$

This illustrates the way SU(3) invariance generates the super-multiplet of mesons shown in Fig. 15.1b.

anti-triplet, $\overline{\mathbf{3}}$, to form a singlet, **1**, and an octet, **8**, may be written symbolically as

$$\mathbf{3} \otimes \overline{\mathbf{3}} = \mathbf{8} \oplus \mathbf{1}. \tag{15.33}$$

By superposing the triangles of Fig. 15.4a on the points of Fig. 15.4a the reader may easily show that the product of two quark triplets produces a $\overline{\mathbf{3}}$ and another triangular pattern of six points, **6**;

$$\mathbf{3} \otimes \mathbf{3} = \mathbf{6} \oplus \overline{\mathbf{3}}. \tag{15.34}$$

We can combine yet another quark triplet, **3**, by superposing appropriate triangles on the points of the **6**, and on the points of the $\overline{\mathbf{3}}$.

The latter combination is already given in (15.33). The combination of **3** and **6** gives **8** and a new triangular decuplet pattern, **10** (see Fig. 15.6.) Collecting these results together we have for the combination of three quarks

$$3 \otimes 3 \otimes 3 = 1 \oplus 2 \times 8 \oplus 10.$$

Since for these, by Table 15.3, $B = \frac{1}{3} + \frac{1}{3} + \frac{1}{3} = 1$ the **8** of this construction satisfactorily accounts for the eight-fold pattern of $B = 1$, spin $\frac{1}{2}\hbar$

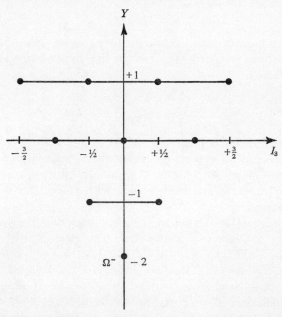

FIG. 15.6. The decuplet, **10**, which may be formed by combining three basic quark triplets, **3**. Nine particles corresponding to the three upper rows, with $B = 1$ and spin $\frac{3}{2}\hbar$, were found by 1962. The tenth particle, Ω^-, corresponding to the lower vertex was discovered in 1964.

particles of Fig. 15.1b. We thus incorporate $\hat{I}_3$ and $\hat{Y}$ into a general (approximate) invariance property of the strong interactions and successfully account in two SU(3) supermultiplets for all the hadrons of Table 11.1.

During 1961 a further octet of particles of spin $\hbar$, was discovered which have $B = 0$ and also fit neatly into the pattern of Fig. 15.5. By 1962 nine other sub-nuclear particles had been found with $B = 1$ and spin $\frac{3}{2}\hbar$. They filled the top three rows of Fig. 15.6, breaking up into an

SU(2) quartet (1238 MeV/c^2) a triplet (1385 MeV/c^2) and a doublet (1535 MeV/c^2) with $Y = +1$, 0, -1, respectively. Invariance with respect to SU(3) then predicted the existence of the missing tenth particle required to make up the decuplet. This was called Ω^-. It has $I_3 = 0$, $Y = -2$ and hence, by (15.28), $Q = -1$. From the masses of the other members of the decuplet its mass could be estimated at 1685 MeV/c^2. The conservation laws of B, Y and I_3 then showed that the easiest way to produce it was with a κ^- beam in a hydrogen bubble chamber, through the reaction

$$\kappa^- p \rightarrow \kappa^+ \kappa^0 \Omega^-.$$

With the estimated mass, it was predicted to be stable for strong interactions, but would decay through the, hypercharge changing, weak interaction

$$\Omega^- \rightarrow \kappa^- \Lambda,$$

or

$$\Omega^- \rightarrow \Xi^0 \pi^-.$$

It was found exactly as predicted in 1964 at Brookhaven Laboratory in the U.S.A., in an experiment using the world's largest existing proton accelerator. This definitely established the approximate invariance of the strong interactions with respect to the SU(3) transformations.

This invariance does not determine the strong interactions uniquely, but greatly restricts the possibilities. The super-multiplet SU(3) patterns, into which the sub-nuclear particles, have been fitted, are rather closely analogous to the periodic table of the chemical elements. They show that the number of different hadrons is not of any fundamental significance. These particles have to be thought of as complicated structures, and there appear to be an unlimited number of them, just as there are an unlimited number of atomic or nuclear energy levels. If they can be thought of as being composed of anything simpler (as the hydrogen levels are all bound states of a proton and an electron) it would appear that they are made of quarks. This is a very revolutionary notion because it involves the existence of particles, with fractional electric charge (in terms of the electron charge) which we have come to regard as an absolute indivisible unit. The existence of three fundamental quarks as a substructure for all the hadrons is a fascinating possibility, which must be tested experimentally in the next decades. However it should be said that the unitary scheme does not logically require their existence. It is possible that all hadrons are "made out of each other" by forces which are approximately invariant for the SU(3) transformations, and that the quarks are just a mathematical device for making calculations. Even so, whether or not quarks exist is one of the most crucial explicit questions now facing fundamental physics.

§ 15.4 Summary

We consider the strong interactions operating between the hadrons. It is found that these can be systematized by attributing to each sub-nuclear particle, three types of charge—the electric charge, Q, the baryon charge, B, and the hypercharge, Y. Each of these is conserved in any strong nuclear interaction and may be related to invariance with respect to a one parameter family of unitary transformations U(1).

The appearance of the hadrons in mass multiplets may be explained by assuming the strong interactions are invariant with respect to the group of two-by-two unitary transformations, SU(2). This gives rise to a new generalized charge, I_3, which is conserved, and replaces Q in strong interactions.

The plots of the observed hadrons in terms of I_3 and Y show regularities which suggest more general invariance properties. These are explained in terms of approximate invariance with respect to three-by-three unitary transformations, SU(3). This incorporates both I_3 and Y conservation and was confirmed by the discovery of the predicted Ω^- particle.

These ideas suggest that all hadrons may be bound states of a triplet of spin $\frac{1}{2}\hbar$ particles called *quarks*, which have fractional electric charge.

APPENDIX

CONSTANTS AND UNITS

A.1 Units of Energy, Momentum and Mass

For atomic physics it is convenient to define as a unit of *energy* the electron volt (eV) which is the energy gained by an electron in a potential of 1 volt.

$$1 \text{ eV} = 1 \cdot 6 \times 10^{-19} \text{ joule} = 1 \cdot 6 \times 10^{-12} \text{ erg}.$$

In nuclear physics the useful energy unit is one million electron volts (MeV)

$$1 \text{ MeV} = 1 \cdot 6 \times 10^{-13} \text{ joule} = 1 \cdot 6 \times 10^{-6} \text{ erg}.$$

The corresponding units of *momentum* are

$$1 \text{ eV}/c = 0 \cdot 5 \times 10^{-27} \text{ mks} = 0 \cdot 5 \times 10^{-22} \text{ cgs}$$
$$1 \text{ MeV}/c = 0 \cdot 5 \times 10^{-21} \text{ mks} = 0 \cdot 5 \times 10^{-16} \text{ cgs}.$$

The unit of *mass* is

$$1 \text{ MeV}/c^2 = 1 \cdot 8 \times 10^{-30} \text{ kg} = 1 \cdot 8 \times 10^{-27} \text{ g} \simeq 2m_e.$$

A.2 Physical Constants

Charge of the electron

$$e = 1 \cdot 6 \times 10^{-19} \text{ coulomb} = 4 \cdot 8 \times 10^{-10} \text{ cgs (esu)}.$$

To avoid introducing electrical units, it is very convenient to define

$$e_M^2 \equiv e^2/(4\pi\epsilon_0) = 2 \cdot 3 \times 10^{-28} \text{ kg m}^3 \text{ sec}^{-2}.$$

Mass of the electron

$$m_e = 0 \cdot 91 \times 10^{-30} \text{ kg} = 0 \cdot 91 \times 10^{-27} \text{ g} = 0 \cdot 51 \text{ MeV}/c^2.$$

Mass of the proton (neutron)

$$m_p(m_n) = 1 \cdot 7 \times 10^{-27} \text{ kg} = 938(939) \text{ MeV}/c^2 = 1836(1838) \ m_e.$$

Planck's constant

$$\hbar = 1 \cdot 05 \times 10^{-34} \text{ joule sec} = 1 \cdot 05 \times 10^{-27} \text{ erg sec} \ [h = 2\pi\hbar].$$

Velocity of light

$$c = 2 \cdot 99 \times 10^8 \text{ m/sec} = 2 \cdot 99 \times 10^{10} \text{ cm/sec}.$$

A.3 Atomic Constants

Radius of lowest Bohr orbit

$$a_0 = \frac{\hbar}{m_e e_M^2} = 0\cdot53 \times 10^{-10}\,\text{m} = 0\cdot53 \times 10^{-8}\,\text{cm}.$$

Fine structure constant

$$\alpha = \frac{e_M^2}{\hbar c} = \frac{1}{137}.$$

Binding energy of hydrogen ground state

$$|E_1| = \frac{1}{2}\frac{m_e e_M^4}{\hbar^2} = \frac{1}{2}\frac{e_M^2}{a_0} = 13\cdot6\,\text{eV}.$$

The corresponding wave number is a Rydberg

$$R_\infty = \frac{|E_1|}{2\pi\hbar c} = 1\cdot1 \times 10^7\,\text{m}^{-1}.$$

The corresponding frequency is

$$R_\infty c = \frac{|E_1|}{2\pi\hbar} = 3\cdot29 \times 10^{15}\,\text{cycles/sec}$$

$$= 3\cdot29 \times 10^9\,\text{Mc/sec}.$$

A.4 Nuclear Constants

Nuclear radius $\sim 10^{-14}$ m.
Nucleon radius (see (9.36)) $\sim 1\cdot5 \times 10^{-15}$ m.
Nucleon Compton radius

$$\frac{\hbar}{m_p c} = 0\cdot2 \times 10^{15}\,\text{m} = 0\cdot2 \times 10^{-13}\,\text{cm}.$$

Nucleon time

$$\frac{\hbar}{m_p c^2} = 7 \times 10^{-25}\,\text{sec}.$$

Cross-section

$$1\,\text{barn} = 10^{-28}\,\text{m}^2 = 10^{-24}\,\text{cm}^2.$$
$$1\,\text{mb} = 10^{-31}\,\text{m}^2 = 10^{-27}\,\text{cm}^2.$$

Binding energy of the deuteron

$$\epsilon = 2\cdot1\,\text{MeV}.$$

INDEX

A

Absorption, 175
Alpha-decay, 93–101
Angular momentum, 9, 55–65, 67, 83, 146
  addition of, 124–125
  commutation relations, 56, 146
  conservation of, 159
  eigenfunctions, 62–64
  eigenvalues, 61, 148
  expansion, 111, 144
  operator treatment of, 55, 146–148
  and spin, 81, (see Spin)
  vector diagram, 61–62
  z-component, 56–57
Anti-particle, 128–129
Anti-symmetry (see Symmetry)
Asymptotic condition for scattering, 109
Average Value, 18, 136
  time dependence of, 155

B

Balmer series, 9
Barn, 104
Barrier penetration, 42, 98
Baryonic charge, 186, 188
Beta decay, 179–183
Binding energy, of deuteron, 118–119, 181
  of hydrogen, 89, 181
Black body radiation, 5
Bohr, N., 3, 9, 10, 25, 75, 177
Bohr levels, 8–10, 24, 25, 32, 34
  recoil correction, 73–75, 89
Bohr megneton, 84
Bohr radius, 10, 32, 71, 181
Boltzman statistics, 86
Bose–Einstein statistics, 85
Boundary condition, 26
  for scattering, 109
Bound states, properties, of, 46

C

Central potential, 67–68
  angular momentum in, 67–68
Centre of mass frame, 114–117
Centre of mass motion, 73–75
Charge; conservation of
  baryonic 186, 188
  electric 186
  hypercharge 187, 188, 193
Classical limit, 157–158
Collision theory (see Scattering)
Commutation relations, 14, 21, 27
  of angular momentum, 81, 146
  of position and momentum, 20, 21
  and uncertainty principle, 27
Commutator, 14, 15
Commuting operators, 30–31
Complementarity, 20
Completeness, 140–144, 149
Completeness condition, 141, 149
Compton effect, 6
Compton wavelength, of electron, 7
  of nucleon, 180, 201
Conservation laws, 159, 160–163
Conservation, of angular momentum, 159
  of parity, 159–160
Constants of motion, 160—163
Continuity conditions, 37, 38, 39, 44
Co-ordinate frames, centre of mass, 114–117
  laboratory, 114–117
  reflection of, 64, 159
Correspondence principle, 19, 31
Coulomb potential, 22, 68
  energy levels, 69–73
  scattering, 93–95, 107–108, 174
  (see also Rutherford scattering)
Cross-section, total, 104, 110

203

PRINTED BY WILLIAM CLOWES & SONS LIMITED, LONDON, COLCHESTER AND BECCLES